# The World
## *of*
# Worms

# The World
# *of*
# Worms

*Dorothy Hinshaw Patent*

HOLIDAY HOUSE · NEW YORK

*Library of Congress Cataloging in Publication Data*
Patent, Dorothy Hinshaw.
  The world of worms.

  Includes index.
  SUMMARY: Discusses a variety of segmented worms, their
characteristics and habits, and their usefulness to humans.
  1. Worms—Juvenile literature. [1. Worms] I. Title.
QL386.P37      595'.1      77-17117
ISBN 0-8234-0319-X

*To bookworms*

# Contents

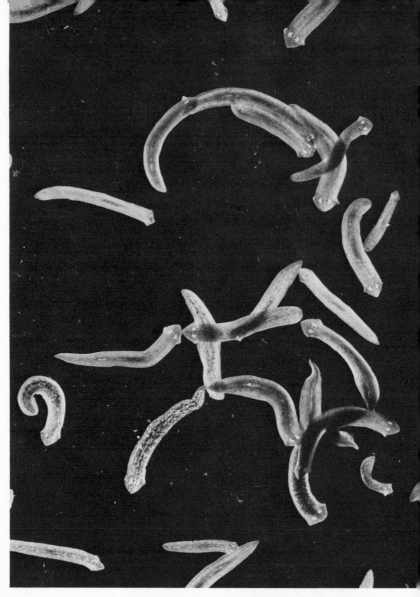

Planarians, shown here enlarged several times, are smoothly gliding flatworms that travel on the hundreds of cilia, or minute hairlike projections, that cover their under surfaces. The two lobes at the sides of the head end are thought to be taste sensors.

# One

# The Simplest Worms

What is a worm? That may seem like an easy question to answer, but really it is not. The first worm to come to mind is the earthworm. And maybe you have heard of the flatworm, or planarian, and the leech. But many other kinds of worms exist as well, some with strange ways of life. Many worms are parasites, living inside the bodies of other animals or plants. Many worms besides the earthworm burrow in the soil, and an amazing variety of worms live in the sea.

The long, thin, legless worm body offers many conveniences to its owner. A thin animal can burrow easily. It can fit into small spaces with little difficulty. All animals need oxygen inside their bodies to live, and they must get rid of wastes such as carbon dioxide. A thin body with a lot of surface area in relation to its body bulk makes this easy. The worm shape also allows an animal to explore with only a small part of its body, its head, while the rest of the animal trails safely behind in safe territory or remains protected within a burrow or crevice.

The success of this body plan has led to many kinds of worm-shaped animals. The hunting proboscis worms, the parasitic tapeworms and flukes, the many tube-dwelling and burrowing marine worms, and the crawling "sea mice" are only a few examples of successful worm life.

The simplest worms belong to the flatworms, or platy-helminthes (*platy* means "flat" in Latin, and *helminthes* means "worms"). We call a group of animals, such as flat-worms, which share a certain body plan, a "phylum." The plural of this originally Latin word is "phyla," and as we will see, there are many different worm phyla. As their name indicates, flatworms have a body which is flattened from top to bottom. The digestive system has only one opening, the mouth, and the nervous system is relatively simple. One thing about flatworms which may seem rather startling at first is that each animal is both a male and a female. Each worm has testes, the male sex organs produc-ing sperm, and ovaries, the female organs which make eggs. This arrangement is actually rather common in the world of invertebrate animals.

### A Trio of Types

There are three different types of flatworms. There are free-living ones such as the common planarian, and there are two groups of parasitic flatworms. Tapeworms are highly specialized parasites, while flukes are more like the nonparasitic flatworms. The free-living flatworms are called turbellarians. While the planarian lives in fresh water, most turbellarians live in the sea. Some have attractive stripes or spots in bright colors. Others are bright green because of one-celled plants called algae which live in or under their skin. A few kinds can swim, but most crawl about in the sand or mud or live on water plants. While the majority of turbellarians are small—no more than ten millimeters long (about ⅓ inch)—a few which live on land in moist tropical areas reach 60 centimeters (almost two feet) in length.

The surface of a turbellarian's body is covered with trillions of microscopic hairlike projections called cilia. These beat with a wavelike motion. Special cells on the body surface produce slimy mucus which covers the bottom of the animal. This wet layer protects the cilia and gives them a fluid to push against as they beat. They are like minute oars, getting a purchase on the mucus as the worm glides along. Several layers of muscles allow these animals to change shape and turn corners. If you have ever seen planarians, you know how smoothly flatworms can move, turning this way and that with ease.

No flatworms have ears, so they live in a silent world. Turbellarians are sensitive to touch and also can "taste" chemicals in their environment. The cells which sense chemicals are scattered all over the body in some kinds, but are more often concentrated in patches on the head. While the eyes of planarians give them an amusing cross-eyed appearance which suggests vision of an advanced sort, turbellarians actually can detect only the presence or absence of light. Their eyes cannot form an image. Since they generally move away from light and live in dark places, these simple eyes meet their needs quite well. Many turbellarians have more than one pair of eyes. There may be two or three pairs or whole patches of eyes dotting areas around the front end. One kind of land planarian has eyes all along the front and sides of its body.

The digestive system of turbellarians is strange by human standards. Some simple kinds actually lack a gut. The mouth leads merely to a mass of digestive cells with no

CAROLINA BIOLOGICAL SUPPLY COMPANY

These planarians are injected and stained to show up internal organs. In the one above the many-branched intestine shows as black, and the tubular pharynx is clearly seen in the middle. In the worm at bottom various organs show at the edges—the yolk glands along the worm's right side, the testes and excretory system along its left. Eyespots at front are rudimentary eyes.

cavity at all. Other turbellarians have a gut which is not the familiar continuous tube but rather a complex branched structure. The mouth is not located at the front but can be found somewhere along the bottom side, often near the head end. A muscular tube called the pharynx connects the stomach with the branched intestine of the larger turbellarians. The pharynx can be pushed out through the mouth for feeding. Some turbellarians capture small animals and swallow them whole, while others push the pharynx into their victims, or into dead animals, and pump out the contents, using the strong pharynx muscles.

### Growing Back Lost Parts

One reason for the planarian's fame is its ability to re-grow lost parts. If a person loses an arm or leg, it is gone for good. But if a planarian loses its head there is no problem. It just grows a new one. Scientists have spent many years studying this ability to regenerate, or grow back, lost parts. They have found that the front part of a planarian re-generates better and more quickly than the rear. They know that different kinds of planarians differ in their ability to regenerate. Some cannot regrow the head at all. Others can, as long as the cut is in front of the pharynx. Still other kinds can be chopped up into many small pieces, and each piece will regrow all its missing parts.

By cutting off a planarian's head and making a slit in the remaining part of the body, a scientist can produce a two-headed planarian. These experiments may sound like un-important fun and games, but from them biologists are learning a great deal about growth and regeneration in animals. Such knowledge can help us understand how

Such is the planarian's ability to regenerate parts that its head end can be split to produce two heads, or even more.

CAROLINA BIOLOGICAL SUPPLY COMPANY

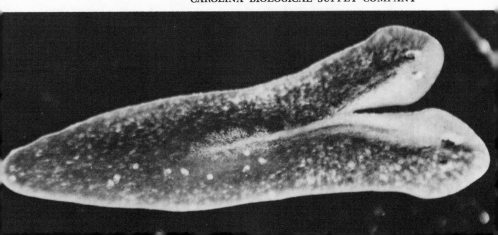

wounds heal. It may someday lead to a way of getting human organ tissues to regenerate, too.

In nature, some planarians reproduce by "falling apart." The worm fragments into several pieces, each of which forms a complete new worm. The familiar freshwater planarian tears itself apart to reproduce. The tail end hangs onto some surface, while the front end crawls away until the worm breaks in two. Then each piece grows back its lost parts to make a complete new individual. There are two worms where there used to be only one. Other turbellarians grow new individuals at the rear without breaking apart. The result is a chain of small worms, each with its own pair of eyes. When one individual is quite well developed, it breaks off and becomes independent.

Even though each turbellarian has both male and female reproductive systems, these worms must mate to reproduce sexually. One worm cannot fertilize its own eggs. When the worms mate, they fertilize one another at the same time. Sperms travel through a long penis and are deposited in a special pouch of the other worm. After mating, the sperms swim through a duct to a spot near the ovaries. There they fertilize the eggs as they pass from the ovaries. Planarians lay their eggs wrapped together in protective cocoons which are attached under stones. In a few weeks the young worms hatch out.

### Flukes

The flukes, or trematodes, are very successful and therefore dangerous parasites. Many kinds of flukes infect people and domestic animals, living in the blood, liver, lungs, intestines, or some other part of the body. They

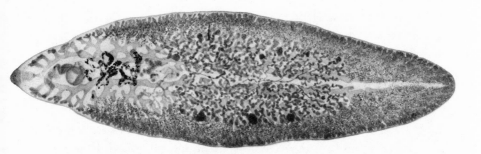

*Fasciola hepatica,* the sheep liver fluke. The mouth is at left, with the branched intestine leading off from it. The testes and excretory system are seen in the middle area, and yolk glands all along both sides, up to the head area. Flukes are dangerous parasites.

sap the strength of their victims and, in extreme infections, may kill them. Along with tapeworms and roundworms, flukes are the most numerous and important parasitic worms.

Life inside the body of another animal is quite different from a free-living life. Flukes are similar in general structure to turbellarians, but they have some important differences which suit them for a parasitic life. They have suckers for hanging on. One sucker surrounds the mouth. This allows the fluke to feed easily. The fluke's body has a covering which lacks the cilia by which turbellarians crawl about. This covering layer protects the worms from being digested by its host if it lives in the intestine. Unlike most other animals, flukes can live and grow in the absence of oxygen.

Many parasites have complex life cycles, requiring more than one kind of host animal. The animal which is the unfortunate host of the adult, reproducing stage of a parasite is called the primary host. Vertebrate animals, especially fish, are usually the primary hosts for flukes. Hosts which are infected by earlier stages of the parasite are called secondary hosts. Some parasites have two or even more secondary hosts.

The Chinese liver fluke of the Orient is one of the most common and serious parasites of humans. The adult flukes live in the human liver. The eggs go from the host's liver, when the female fluke lays them, by way of the bile duct into the intestine. They pass from the body in the host's feces. If liver fluke eggs reach the water, they hatch into swimming larvas with many cilia. The larvas swim around until they encounter a particular kind of snail. Then they bore into the snail's body.

Sometimes the snail may eat fluke eggs before they hatch. Then the larvas hatch in the snail's gut and bore through into the body. Each larva grows inside the snail and changes into a second type of larva which has a hollow body. Inside this body, nourished by the snail's tissues, little bunches of cells form. Each bunch then grows into still a third type of larva. These in turn become hollow and develop a whole mass of yet another larval stage. This fourth larva is a more nearly complete animal called a cercaria. It has a digestive system, a sucker, and a tail. The cercarias leave the snail and swim about. If they find a fish in their wanderings, they burrow in and form pro-tected cysts in the fish's muscles. If a human being eats such a fish raw or undercooked, the cysts are digested away by his digestive enzymes. The cercarias which are released then penetrate his blood vessels and travel to his liver, where they grow to adulthood.

This life cycle seems very strange. It is hard to believe that it exists and is successful. But a little thought can make some sense of it. The chance of one larva finding a snail may be small. But the adult fluke lays thousands and thouands of eggs. And in wet farm areas where indoor plumbing is lacking or in cities where raw sewage is released into ponds and streams, the chances are good that some eggs will reach a pond where snails live.

Once inside the snail, the egg becomes many individuals instead of one. Let us say that 20 third-stage larvas grow inside one second-stage individual. Then each of these grows 20 cercarias inside. Four hundred cercarias would swim out of that snail from just one egg. The chances are that many of these will find their way to fish. The secret of success of the Chinese liver fluke, however, lies in the eating habits of humans. If everyone cooked fish well before eating it, all the encysted flukes would die and none could infect people. But in the Orient, where these animals live, raw fish is considered a delicacy. As long as people there eat raw fish, it will be very hard to get rid of the Chinese liver fluke.

Some flukes increase their chances of reaching their primary host by the ways they affect the intermediate host. Some fish parasites encyst in the eyes, making their hosts blind and therefore more vulnerable to primary-host predators. Another kind, which infects stickleback fish, slows the fish down so much that they are easy to catch. A different fluke causes its fish host to darken in color, making it more obvious to predators. Flukes may also affect snail hosts. Some infected snails grow larger than normal and stop hiding, making them easy marks for unwitting primary-host predators.

## *Tapeworms*

While flukes resemble turbellarians in many ways, tapeworms are so completely modified as parasites that they appear quite different. Like flukes, tapeworms have a protective outer layer, or cuticle. They have a strange, small "head" called a scolex. The scolex has suckers and hooks by which the tapeworm hangs on inside the intestine of its host. All tapeworms live in the digestive tract of vertebrate animals. Here they are surrounded by digested food. So perhaps it should not surprise one too much to learn that a tapeworm has no digestive system of its own. It simply absorbs the food its host has already digested and never needs actually to feed.

The body of an adult tapeworm consists of the scolex and a long chain of body sections. New sections are budded off behind the scolex and mature as they grow. Each section contains complete male and female reproductive systems. In most tapeworms, the male system matures first in one section. The female system of that section matures later. Thus a younger male segment can fertilize an older female segment of the same worm. But when two worms are infecting the same host, one worm can fertilize the other. The fertilized eggs are coated with several protective layers. Embryos develop in the eggs while they are still inside the worm's body. The other parts of the body section disappear as the embryos develop, and soon it becomes merely a sac full of protected embryos within their eggs. These mature sections break off at the rear of the animal and pass out with the feces.

### Tapeworms in Humans

Like flukes, tapeworms require at least one intermediate host to complete their life cycles. The human being is a primary host for several different tapeworms, with fish, pigs, and cows as secondary hosts. The pork tapeworm is perhaps the best known of these. This worm can reach a length of three meters (about ten feet) in the human intestine. The body sections are released into the feces. The eggs hatch only if eaten by an intermediate host such as a pig.

When an egg hatches, a round larva with minute hooks on it is released. This larva bores into the wall of the pig's intestine and passes into the bloodstream. The blood carries it to body muscles where it develops into a peculiar larva called a bladder worm. This is a rounded sac about ten millimeters (approximately ⅕ inch) in size. It contains the scolex of the worm. The scolex is inside-out. If a person eats raw or undercooked pork containing the bladder worm, the scolex turns right side out inside the intestine and hooks on. There it grows into an adult worm.

### Ribbon Worms

It's hard to believe in a worm some 30 meters (over 90 feet) long. Some scientists do not believe it either, but that length has been claimed for the soft, blackish bootlace worm of the North Sea. These worms are hard to measure, for their bodies are quite elastic and can be stretched out to great lengths. But the collector must be careful, for bootlace worms may also fragment into as many as forty

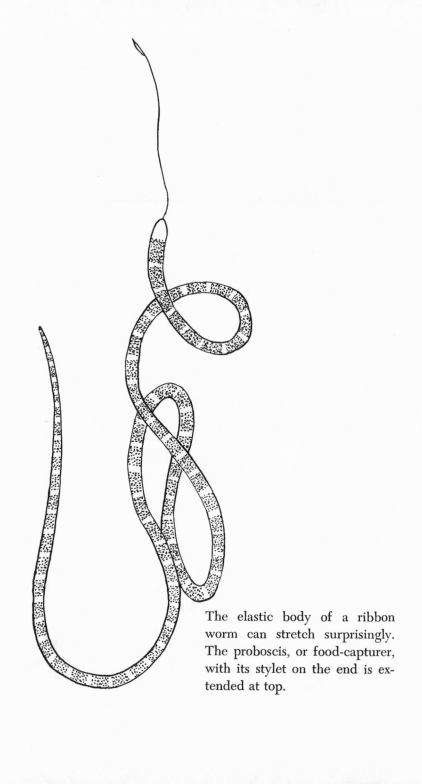

The elastic body of a ribbon worm can stretch surprisingly. The proboscis, or food-capturer, with its stylet on the end is extended at top.

pieces, making the task of measurement quite impossible. These peculiar creatures are members of the small but fascinating worm phylum Nemertea, the ribbon worms or proboscis worms. The name "ribbon worm" comes from their long, usually flattened shape, while "proboscis worms" refers to their unique and amazing food-capturing device.

In a special fluid-filled cavity above the digestive tract of a nemertean lies a long, hollow tube called the proboscis. It extends far back into the body of the animal, finally ending with muscles attaching it to the blind end of the cavity. When the worm senses prey, its muscles contract violently. This puts pressure on the fluid in the proboscis cavity, forcing the proboscis to evert, or turn inside out. It rolls out of the cavity with lightning speed and whips around the worm's prey, capturing it firmly. The victim is then dragged back to the mouth and eaten. A small creature can be consumed in this way in less than a half minute.

The proboscis never turns completely inside out. The muscles at the far end always hold part of it back. These muscles contract to pull the proboscis into its cavity again when it is not being used. Some nemerteans have sharp stylets on the proboscis. These are located part of the way back, right at the point which will become the tip when the proboscis is everted. The force which pushes out the proboscis is strong enough to drive the stylet into the prey's body like a dart. It may be a poison dart, for some nemerteans have poison glands leading to the stylet. The proboscis may also come in handy for defense, or it can be used as a digging tool by some ribbon worms.

In most of their body structure, ribbon worms are similar to flatworms. Aside from the proboscis cavity and gut, the nemertean body is solid like that of flatworms. Most have cilia and mucus glands on the ventral side and glide along as do flatworms, but swimming kinds can wriggle from side to side by using the body muscles. The nervous system of nemerteans is only slightly more complicated than that of flatworms, and they have no real head to speak of. They do have a true circulatory system, however, with a rather sloppy, inconsistent blood flow. The digestive system is more advanced than that of flatworms, for nemerteans have an anus, an opening at the far end of the gut. This allows for a one-way system with specialized parts such as an esophagus, stomach, and intestine.

Like flatworms, nemerteans have great powers of regeneration, and some kinds like the bootlace worm may fragment into many pieces. Each piece can grow into a complete new worm. Males and females are separate in ribbon worms. The eggs and sperms are released into the sea water where fertilization takes place.

While some nemerteans are very long, most are of modest size, around 20 centimeters (some eight inches) in length. They may be brown or black, but often have bright shades of red, orange, blue, or green instead. Ribbon worms are common along both coasts of the United States but are generally hard to find since they hide during the day. Some kinds live in the deep sea, while others float with the currents. But all use their unique proboscis as a deadly hunting weapon.

# Two

# Roundworms Are Everywhere

Roundworms, or nematodes, are among the most numerous animals on earth. Roundworms have been thawed out alive from Antarctic ice and scooped from steaming hot springs. They thrive high in the mountains and at the depths of the ocean. Some kinds are found in soil and water throughout the world. Others can survive in only a very limited habitat—one kind lives only in impure vinegar, while another is found in the beer-soaked mats of German beer halls. While most are free-living, many kinds are parasites of plants and animals. One scientist remarked that if everything else on earth magically disappeared, we could still see the forms of both the living and nonliving worlds etched in roundworms. Another patient scientist took apart a decomposing apple from an orchard and counted 90,000 roundworms of various kinds living in it. A square meter of sea mud may contain over four million of these wriggly creatures, while billions make their home in an acre of good farm soil.

If these creatures are so incredibly abundant, why have most people never seen them? Most-free-living nematodes are very small, less than one millimeter long. To see them requires the use of a hand lens or microscope. Many

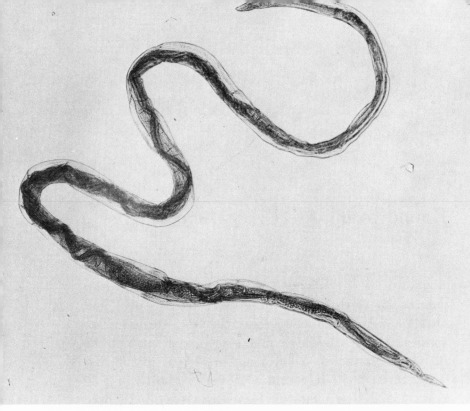

Many roundworms are parasitic. This one is called the stomach worm (*Haemonchus contortus*); it can infest the stomachs of mammals and birds.

parasitic nematodes are much larger, but they spend most of their lives hidden inside the bodies of other animals or plants. The horse nematode can be 35 centimeters (about 14 inches) long, and the female of the parasitic guinea worm can reach a lenght of 120 centimeters (some four feet). Other nematode parasites are very small, such as many plant parasites which infect roots. The African eye worm, which may find itself wriggling across a human eyeball, is only a few millimeters long, as are the filaria worms which are carried to their vertebrate hosts by bloodsucking insects.

## A Common Body Plan

While there are thousands of kinds of roundworms, the basic body plan is varied only slightly throughout the phylum. *Nematode* means "threadlike" in Greek, for these worms have long thin bodies with tapering ends. This shape leads to one descriptive name commonly used —eelworms. As their other common name, roundworm, indicates, their bodies are round in cross section, not flattened like those of flatworms. Nematodes are easy to recognize since almost all have this same basic shape. Nematodes have a mouth at the front end of the body. The anus is near the tail end. A protective cuticle covers the body.

Inside, a straight digestive tract runs from the mouth to the anus. The gut lies inside a fluid-filled body cavity which gives support to the worm. While they lack respiratory and circulatory systems, nematodes do have a

Nematodes are long and narrow; almost all have essentially the same shape. This is a pinworm, which can enter the stomach of various vertebrates. A bad infestation is especially troublesome in children, causing loss of weight, extreme tiredness, irritability, fever, and other symptoms.

CAROLINA BIOLOGICAL SUPPLY COMPANY

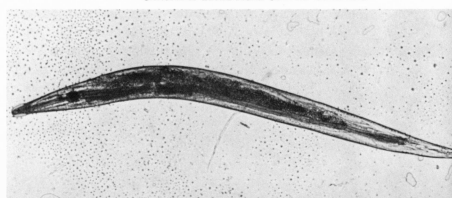

peculiar sort of excretory system. Their nervous system has a nerve ring near the front and four nerve cords passing backward through the body. While a few nematodes have a pair of simple eyes, most are blind. The skin of most nematodes, however, appears to be sensitive to light. Nematodes are also sensitive to touch and to chemicals in their environment.

The only muscles of roundworms are longitudinal, running from the front to the rear of the animal. There are two sets of these muscles along the bottom side and two sets along the back. When the bottom sets of muscles contract, the front and rear of the worm bend downward. When the sets along the back contract, the worm bends the opposite way. Because it has only these muscles, a nematode cannot bend from side to side. When it moves, a roundworm uses these bends of the body to push against the soil particles or sand grains. A wave of contraction of the back muscles is followed in turn by one in the bottom muscles, pushing the body through the soil. The worm looks much like a moving snake, except that the snake is bending from side to side instead of from top to bottom.

If nematodes are removed from the soil and placed in water, their bodies have nothing to push against. They just thrash helplessly about. In order to move efficiently in one direction, a roundworm needs something to push against; it cannot swim well. For this reason they are rarely found living free in the water. They make their homes on the bottom instead, where they can move with ease. It may seem strange that an animal which can barely swim, if at all, can be active only in a wet environment. But all nematodes require a watery habitat. Those which live in the soil survive in the film of water surrounding soil particles. Since the insides of plants and animals are very moist, parasitic nematodes have no problem of drying out once they have found a host.

## Different Ways of Feeding

The one part of the roundworm body which shows great variation is the mouth. Nematodes feed in every imaginable way. Some eat minute particles of food found in the soil. Others are predators which grab prey, such as other nematodes, and suck out their insides. Plant parasites often suck plant juices, while animal parasites may feed on the contents of the host's intestine or drink its blood. The region of the mouth is changed in different ways from species to species, making these varied diets possible.

Most free-living nematodes have six lips surrounding the mouth. Parasites may have fewer lips, and in many roundworms the lips have spines, hooks, or suckers associated with feeding. The mouth cavity itself may also be modified. In worms which feed on bacteria or fluids, it may be simply a narrow tube through which food is sucked. Many nematodes have a sharp, pointed stylet which is used to pierce plant cells or animal prey. The stylet can be pushed out of the mouth by muscles and stuck into the host. Digestive enzymes are then injected through the stylet, and the partially digested food is sucked back into the worm's gut. Many predatory nematodes have powerful jaws used for biting prey. They may have teeth for holding their victims and biting off pieces, or they may swallow them whole.

Nematodes which are animal parasites feed in various ways. Those that merely eat the partially digested food

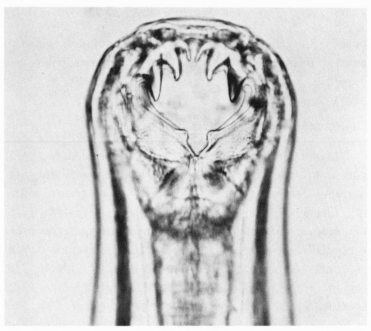

The hookworm, like many other parasitic nematodes, has hooks with which it attaches itself to the intestinal lining.

in the host's intestine, such as Ascaris, have a simple mouth with no teeth or jaws. The food is simply sucked in by the powerful muscular pharynx. Hookworms, on the other hand, feed partly on the cells lining the intestine. They have a large mouth cavity with hooks at the opening that allow it to grab the host's intestine. A chunk of tissue is pulled into the wide mouth. Cutting teeth inside scrape off bits which digestive juices act upon. The resulting soup is sucked in by the pharynx and swallowed. Some hookworms prefer to feed on blood, rapidly pumping large quantities of it through their bodies almost unchanged until it becomes part of their tissues.

## Life Histories

While the life styles of roudworms are extremely varied, all go through the same basic stages. Most species have separate sexes. The smaller male has special spines which aid in mating. The eggs are fertilized inside the female and are then covered by a thick shell. When the egg hatches, a small worm comes out. Almost all nematodes have four immature stages. These stages are very similar to the adults, except that they are smaller and lack parts of the reproductive system. The young worms molt four times, shedding the cuticle and growing a new one at each molt. The adult worms do not molt, but they may continue to grow. While most nematodes do not get much bigger as adults, the big animal parasites such as Ascaris and guinea worms do most of their growing after the last molt.

Millions of nematodes live freely, without depending on a plant or animal host to survive. Hundreds of kinds live in muddy ocean bottoms. Others are found in beach sand and lake bottoms, while some can survive the very high temperatures of hot springs. Soil nematodes are very important to the balance of nature in the earth, and many kinds are important in the breakdown of dead organic matter. But from a human point of view, the most important roundworms are the parasitic ones. These may cause serious damage to important crops or injury and loss of vitality to domestic animals or to people. For this reason, scientists have concentrated their energy on nematode parasites, and we know more about them than about the kinds which are not dangerous to us.

## Plant Parasites

Many important crops from onions to apple trees are threatened by nematode parasites. In California alone it is estimated that these animals cause over $90,000,000 worth of crop damage each year. Many kinds of worms are responsible and cause different symptoms. Infected plants may grow more slowly and yield less produce. Or the plants may just wither away and die. Some nematodes cause growth distortion in plants, resulting in strangely shaped leaves or lumpy tumors called galls on roots or stems.

One of the most damaging plant parasites is the stem eelworm. Because it infests underground bulbs of onions and other plants, it is also called the bulb eelworm. This very versatile worm infests plants from almost 50 plant families which include many economically important crops. It is found in many parts of the world from Pakistan to the United States but is especially damaging in the cooler, moister parts of western Europe.

Like many other nematodes, the stem eelworm can survive a variety of harsh conditions which would kill off less hardy creatures. It can live through subzero weather or temperatures as high as 55° C (131° F). One of its startling secrets of success is the ability to survive drying. The fourth-stage larva can curl up and dry out but not die. It is of course inactive, but when exposed again to water, the worm swells back to normal size and shape and wriggles away. Certain chemicals found in the cuticle only during this fourth larval stage probably help protect the worm when it is exposed to such harsh conditions. The ability to

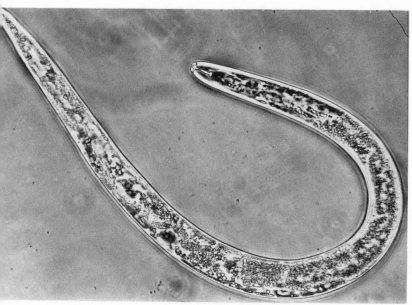

Nematodes cost farmers millions of dollars in damage to crops. This plant nematode is *Pratylenchus penetrans*, much enlarged, which feeds on plants roots and also lay eggs in them; the result is death for the plant.

dry up and survive aids in the spread of this destructive worm. It can be carried on seed exported to different countries, and it probably also travels on wind-blown seeds, such as those of the false dandelion in California.

As one might suspect, the hardy fourth-stage larva is the one which infects plants. After a rain, these worms are found at the soil surface. They appear to be attracted by certain chemicals produced by host plants and usually attack at or above the soil surface. The infecting larva probes the plant with its sharp stylet. When it finds a weak place it pierces the plant and enters. Enzymes in the worm's saliva dissolve away part of the rigid plant cell walls which give them shape. The cells, which were elongated and cemented together, providing structure to the plant, separate from one another and round up. The plant becomes soft and mushy.

Although it is called the stem or bulb eelworm, this unpleasant little organism lives in all parts of the host plant. It may infest the leaves and flowers, carried to different parts as the plant grows. When they have successfully invaded a plant, stem eelworms reproduce very quickly. Females must mate with males before egg-laying, but they can begin to lay as early as three days after the final molt. The eggs are put right inside the plant and hatch as second-stage larvas. These become adults in only nine to 11 days, making the complete life cycle only about three weeks long. Since each female can lay between 200 and 400 eggs during her short life, the population of worms in a plant can grow to a large size very quickly.

The life span of a stem eelworm depends on environmental conditions. If it is inactive and dried up, it may survive for more than 20 years, awaiting the vital moisture it needs to lead an active life. If it lives in soil without host plants, it can hold out for as long as a year and a half. But if it invades a host plant and reproduces, it may become old and die in less than seven weeks.

How can farmers protect themselves from such hardy, well-adapted enemies? One successful strategy involves developing resistant crops. Resistant strains of rye, oats, and red clover are very successful. Stem eelworms can invade and survive in the resistant plants, but cannot develop properly or reproduce there. Heat can also be used to kill the worms. If strawberry plants are submerged in water at 46.1° C (115° F) for seven to eight minutes, nematodes die while almost all the plants survive undamaged.

Other methods can control plant parasitic nematodes as well. Crop rotation is the simplest, but requires knowledge of the pest's life style. Since many nematodes can survive in the soil for more than a year without a host, crops must be rotated on a three- to six-year schedule. This means that a crop planted one year should not be planted in the same field for the next few years, depending on the crop and the pest. The farmer must also be careful not to plant a different crop which is also a host for the pest he wants to eliminate.

Organic fertilizers also lead to reduction of parasitic nematodes in the soil. Manure or chopped up cabbage leaves work well, as do other plant and animal wastes. And finally, some farmers use poisonous chemicals to kill off the worms. Since poisons generally affect more than just the dangerous animals, other methods of control are better, when it is possible to use them.

### Threats to Humans and Their Animals

A dizzying variety of roundworms infect humans and other mammals. While wild animals are common hosts, domestic ones unfortunately have their share, too. Cows, horses, pigs, goats, sheep, chickens and other domestic birds are affected. Millions of dollars are lost by farmers each year in this country alone because of these worms, and many animals die painful deaths. Humans, especially in tropical, underdeveloped countries, are hardly better off than their stock. Roundworms of various kinds attack the human digestive system, lungs, liver, skin, lymph system, kidneys, blood vessels, and eyes.

Some roundworms have alarming effects. The huge guinea worm is found mainly in southern Asia, Africa, New Guinea, and parts of Russia. The females may be 70 centimeters (about 28 inches) long and live under the skin when they are mature. They usually are found on the feet or legs and release their larvas through a hole in the skin when the infected person steps into the water. To get rid of the worm, it must be caught while part way out of the hole. Then it is very slowly wound up on a stick. It may take weeks to do this, and the process cannot be hurried for fear of breaking the worm. A dead, broken worm can cause serious infection in the unfortunate host. Can you imagine walking around for weeks with a live worm wound around a stick on your leg? While some drugs can help speed up the winding process, this primitive method is still the most popular way of removing these strange worms.

Elephantiasis is another alarming nematode disease. Long, thin worms called filarias infect the lymph system. (This system has vessels which contain body fluids; it empties into the circulatory system.) At first there are no symptoms. Then, when the worms block the lymph vessels, fluid accumulates. The victim suffers swelling of the arms and legs with pain, numbness, and weakness. These symptoms come and go over a period of weeks or months and are followed by recovery. But if the person keeps getting reinfected over a period of years, permanent damage to the lymph system can result. The lymph vessels break down and enormous swelling of the leg or other affected part can result. This swelling is called elephantiasis and fortunately affects only a small number of people in the regions where these filarial worms are common.

## The Parasite Ascaris

Different kinds of Ascaris infect humans, pigs, cows, dogs, cats, and poultry. These especially large roundworms are studied in biology class because of their size and abundance. This pig roundworm is the most important swine parasite there is. Farmers in America lose around $35,000,000 each year because of these worms. They are the biggest cause of death and damage to pigs raised under unsanitary conditions.

The pig roundworm has a direct life cycle with no intermediate hosts. It is incredibly fertile, with each female producing one to two million eggs a day. The eggs are produced literally as fast as bullets shot from the barrel of an automatic rifle. The females can store up enough sperm to fertilize their eggs for at least sixteen weeks even if no males are present.

The human Ascaris worm is also a common parasite. Females of this kind are 20 to 40 centimeters (about 8 to 16 inches) long, while males are a bit smaller. They too have a direct life cycle. The eggs are passed out in the feces. They develop in the soil until the second larval stage. Then, if they are eaten, they hatch in the intestine. The larvas burrow through the intestine wall into the lymph system or into the smallest veins. From there they are carried to the liver, where they live and grow for about a week. Then they move on with the blood through the heart to the lungs. They leave the blood system in the lungs and develop in tissue to the fourth larval stage. They then burrow out of the tissues into the air spaces of the lungs. They are coughed up into the esophagus and swallowed.

After reaching the intestine, the worms settle down for the rest of their lives. They molt to the final adult stage in about four weeks and begin producing eggs two months or so after first entering the body. The adults live only a year or two, but since a female produces around 200,000 eggs each day, she lays 26 to 27 million eggs in her lifetime. These eggs are very tough. They can survive at least six years in the soil and are resistant to almost all poisons.

While a mild infection with Ascaris may cause no serious symptoms, heavy infections can be dangerous. They may cause pain, nausea, fever, weight loss, and other symptoms. Children are more likely to be seriously affected than are adults. The infection may cause the liver or other organs to be inflamed, and if worms accidentally reach the brain, it can be dangerous indeed. While the worms are in the lungs, pneumonia may result, especially in children. The best treatment for Ascaris is of course prevention, with good sanitation being very important. Adult worms can be eliminated easily by worming medicines, but the symptoms caused by the wandering larvas are hard to treat.

### The Threat of Hookworms

Hookworms present a very serious human health problem. They are probably the most common infection of humans in the moist tropics and subtropics. Over 600 million persons may be infected with these unwanted guests. Although they are less than an inch long, hookworms have serious effects on their hosts because of the blood loss they cause. In ancient China hookworm disease was called the "Able-to-eat-but-lazy-to-work yellow disease." "Yellow" probably referred to the anemia which is the major symptom of hookworm infection. So many people lose so much blood from these nasty parasites that one scientist calculated a blood loss each day throughout the world equal to all the blood in the bodies of one and a half million people.

Adult hookworms live in the intestine, attacking its walls with their sharp teeth. Each female hookworm produces thousands of eggs each day which leave the body in the feces. The larvas then develop outside a host to the third larval stage. The larvas then wriggle their way up onto plants. If they reach a passing barefoot person, they grab the skin and burrow into the lymph vessels or the blood vessels. They are then carried through the heart to the lungs. There they burrow out and are coughed up to the

Dog hookworms, *Ancylostoma caninum*; like other hookworms, they feed on the intestinal wall (note the hooks, lower left) and cause lethargy, retardation of growth and of resistance to diseases.          CAROLINA BIOLOGICAL SUPPLY COMPANY

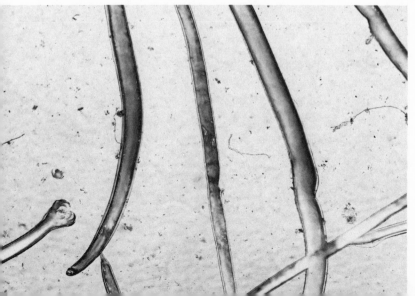

esophagus and are swallowed. In the intestine they settle down and grow to adulthood. They mature quite rapidly, for eggs begin to appear within about six weeks after the worms burrow into the skin.

People with hookworm disease, especially children, suffer from loss of blood. When feeding, hookworms can suck 200 times a minute. They extract the blood sugar and some other nutrients, but leave most of the blood untouched. It passes through their digestive systems in only a minute or two. This may seem wasteful, but to the hookworm it is fine. There is plenty of blood in each human being to feed the worms. But the human may suffer serious symptoms.

Blood is important because it contains a chemical called hemoglobin. Hemoglobin carries oxygen throughout the body. The body cells need oxygen to work well. If lots of hemoglobin is lost through bleeding, the body cannot get enough oxygen to function properly, and the person feels weak and tired. Since hemoglobin contains iron, someone who loses blood must have plenty of iron so his body can replace the lost hemoglobin. If persons with hookworm disease eat foods with lots of iron, they are not very sick. But if they do not get enough iron, they may become seriously ill. Their hearts may become enlarged, and they are likely to lose weight as well as become breathless and tired. Iron is found mostly in meat and fresh vegetables. Since poor people often eat little of these foods, they are likely to suffer badly from hookworm disease if they get it. And because they may not be able to afford shoes, they are quite likely to become infected with hookworms through their bare feet. As long as so many of the world's people are poor and live where sanitation is not good, hookworm disease will be all too common a health problem.

## Always Cook Pork Thoroughly

You probably know that pork should be eaten only when "well done." But do you know why? The reason is a roundworm less than a half centimeter ($\frac{2}{10}$ inch) long called the trichina, or the garbage worm—scientifically, *Trichinella spiralis*. It causes a disease called trichinosis, usually a result of eating undercooked pork. While most parasites are particular about the hosts they infect, the trichina worm is very versatile. It may attack humans, pigs, dogs, cats, rats, and horses as well as dozens of wildlife hosts. When a parasite infects so many kinds of animals, it is difficult to control. Even if it could be eliminated from domestic animals, reinfection by way of wild hosts would always be possible. Occasionally trichinosis is caused by eating some other meat than pork—bear meat, for instance.

Most parasites are common in the tropics. But trichinosis is a disease of cooler climates. It probably originated in the northern regions. One scientist has proposed that it began as a parasite of walruses and polar bears. The infective trichina larvas are found in meat. Since polar bears can kill walruses on land and eat them and walruses can kill polar bears in the water and eat them, it seems a natural beginning for the vicious cycle of trichinosis. From their original hosts, trichina worms infected other wild creatures such as wolves and wild pigs and eventually found their way to people.

Since the infection is passed through eating meat,

humans are usually a dead end from the worm's point of view. But that does not make the disease any less dangerous for the unfortunate person who may have it. Eskimos eat raw meat and have frequent outbreaks of "meat poisoning," probably caused by the trichina. The disease is also common among people such as Germans who eat raw pork sausages. On the other hand, trichina infection of humans is not found in countries like Israel where people do not eat pork.

Trichina larvas settle in the muscles of their host. The most common sites are in the diaphragm, tongue, and chewing muscles. They grow there for a couple of weeks and soon the muscle forms cysts around the larvas. These may live in the muscles, waiting (so to speak) to be eaten, for as long as 11 years. But they do not develop any further unless they are eaten by another animal.

If they are eaten, the cysts are digested away and the larvas released. They enter the wall of the intestine, molt, and begin breeding in only a few days. The larvas develop in the body of the female. When they are ready to be born, she deposits them in the lymph vessels, and they are on their way. Once they leave the lymph vessels and enter the blood, they have a long journey. They pass through the heart to the lungs and back to the heart again. Then they are pumped with the blood throughout the body. When they reach the muscles, they burrow in.

If a person is unfortunate enough to eat infected pork which has not cooked long enough to kill the larvas, he or she may be in for big trouble. A mild infection may have few consequences, but the larvas can be incredibly dense in heavily infected muscle. There may be hundreds or even thousands of cysts in one gram (about $\frac{1}{25}$ of an

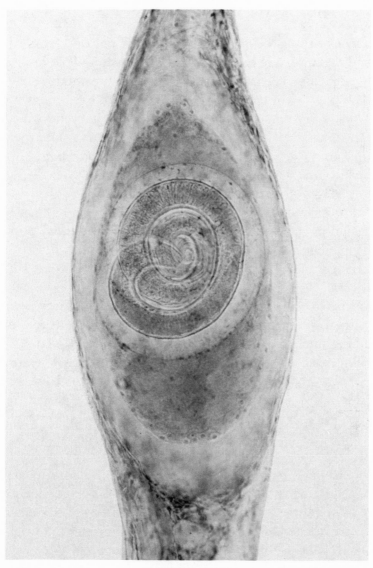

A Trichinella larva, or garbage worm, encysted in muscle. These can cause serious disease in humans and a variety of other animals.

ounce) of meat. The victim may suffer diarrhea, nausea, and cramps while the adults are in the intestine, followed by difficulty chewing, swallowing, and breathing a few days later when the larvas are burrowing into the muscles. Then, a few weeks later when the larvas are encysting, the infected person may suffer from painful swelling of the arms, legs, abdomen, and face, accompanied by pneumonia, anemia, and fever. Some people even die at this stage.

Fortunately, trichinosis is much less common in people than it used to be. There are several reasons for this. Most people know that pork must be thoroughly cooked. Ordinary curing of pork by salting and smoking will not kill trichina worms. But luckily for us, freezing does. Since meat is so commonly quick-frozen and stored that way now in the United States, trichina worms rarely make it to the human intestine alive.

Pigs usually get this infection from eating meat scraps in raw garbage. In many countries such as Canada, there have been laws for years requiring that garbage fed to pigs be cooked. These laws were designed to protect people from trichinosis. But here in the United States, cooking garbage became law only in the 1950s when a serious pig disease which does not affect people broke out in California and spread throughout the country. Quickly, most states passed laws requiring that garbage be cooked before it is fed to pigs. Concern over loss of pigs did what concern for people could not do.

# *Three*
# Many Ways of Life

The most varied and fascinating worms are the annelids, those with segmented bodies. These worms come in a great variety of shapes and sizes, from microscopic thin wrigglers to giant earthworms and fat, crawling sea worms. Some annelids are smooth and slippery like leeches, while others bristle with sharp, poisonous spines, like fireworms. There are colorful annelids—leeches with bright, complicated patterns, and fanworms with vibrant red or green crowns, as well as dull brown earthworms and transparent water worms. This phylum includes leeches, earthworms, and marine worms called polychaetes. Perhaps you have seen a film of sea life showing fanworms (also called feather-duster worms) or bristly bright fireworms. Both of these and many more worms are polychaetes.

Annelid worms are very useful to scientists as research animals. Many kinds, such as earthworms and many tube-dwellers, are easy to obtain in large numbers and simple to maintain in the laboratory. While annelids have been used for many kinds of research, including study of regeneration, embryology, immunity, and blood pigments, perhaps their most important use is in the study of the nervous system.

The annelid nervous system is much simpler and easier to understand than the complex one of a vertebrate animal. And many annelids have quite large nerves which are easy for scientists to work with. These "giant fibers" conduct messages especially rapidly from one end of the worm to the other and allow it to pull away from possible danger almost instantly. They are particularly well developed in annelids which burrow or live in tubes. Because of the large size of these nerves, scientists can insert very fine instruments right inside the cells to measure electrical and chemical changes in the nerves which accompany their functioning. Such studies help us to understand how nerves in general work.

Annelids share a common body arrangement, or "plan," which has proved successful for many ways of life. They show how adaptable the worm-shaped body is, with crawling, swimming, and burrowing kinds. Leeches are parasites and hunters which live in water or in moist tropical jungles; they can crawl inchworm style and in some cases swim. Earthworms are familar burrowers, known to everyone. Some small earthworm relatives are common in ponds as well, gliding among the plants or burrowing in the mud. But the polychaetes show the greatest variety of annelid life styles. Some spend their whole lives in the sea bottom, while others float endlessly with the ocean currents. Many kinds hide in burrows or under stones during the day and hunt among the rocks or reefs at night for food. Some polychaetes have large eyes and powerful jaws that help in capturing strong prey. Others lack eyes and jaws altogether and spend their lives collecting food particles with sticky tentacles or feathery crowns.

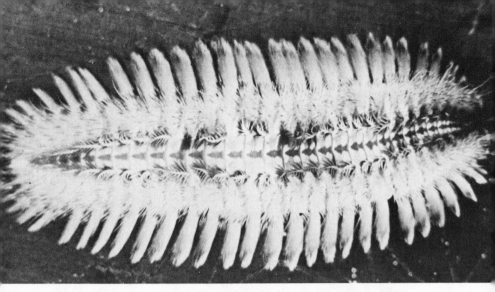

This bristle worm, *Chloeia viridis*, shows its segments clearly, with one pair of bristly parapods extending from each segment.

## The Annelid Body

Flatworms and ribbon worms have solid bodies filled with cells. Roundworms have a large, unique type of fluid-filled space inside. The gut lies loosely within it. This cavity has no special lining. But annelids have a body cavity lined on all sides by a thin sheet of cells. The gut is suspended inside the body cavity by parts of the cell sheet called mesenteries. The main blood vessel, which runs above the gut, and the nerve cord, which is located below it, are also surrounded and supported by these mesenteries. This kind of body cavity, which is lined on all sides by such sheets of cells, is called a coelom. It is a feature of advanced types of animals such as snails, insects, starfish, and vertebrates as well as of annelid worms.

The annelid body consists of a head followed by many body sections called segments. One can see these segments easily by looking at an earthworm. The earthworm looks as if it were made up of a string of rings glued together; each slight bulge represents one segment. The word "annelid" actually means "the ringed one." Each segment contains branches of the nervous and circulatory systems. Each has a pair of funnel-shaped organs called nephridia which eliminate chemical wastes from the body fluids. Each segment contains two compartments of the coelom, one on the right side and one on the left. The two compartments are separated by the mesenteries, which support the gut, nerve cord, and main blood vessel. At the front and rear end of each segment is a sheet of tissue called a septum. The septa separate each segment from those in front of and behind it.

Why should annelids have a body divided up into so many small compartments? What use is it, anyway? The answer is that a body with individual segments, each with its own muscles and fluid-filled compartments, can move in many ways. Roundworms can bend only from top to bottom; they cannot move from side to side. They cannot lengthen or shorten their bodies either. But annelids can become long and thin or short and fat. They can make some segments thin and others fat. They can bend from side to side or up and down. The muscles of the annelid body push against the fluid in the compartments. If the circular muscles which run around the body like rings contract, the segment becomes longer and thinner. If the longitudinal muscles which are attached at the front and back ends of the segment contract, the segment becomes shorter and fatter.

This type of movement is especially useful in burrowing. A burrowing worm can anchor itself in the ground with some short fat segments while it stretches out to explore around with its front segments. It can escape from enemies by quickly contracting its longitudinal muscles, shortening and pulling in its body in a flash. It can easily twist and bend it way around rocks in the sail. The annelid body is probably the most flexible one in the whole animal world.

### How Polychaetes Are Put Together

While polychaetes form a more varied group of animals than leeches or earthworms, their body plan shows the basic annelid characteristics most clearly. Nereis, the clamworm, is a typical unspecialized polychaete. Its body is adapted for crawling, swimming, and burrowing. It can do all of these quite well. Later we will see how other polychaetes have become "experts" at one of these three kinds of lives at the expense of the other two.

Nereis burrows in the mud during the day but some kinds venture out at night to feed. There are several kinds of nereids. All have a strong pharynx which can be pushed out of the mouth. A pair of heavy jaws at the end of the pharynx are used in feeding. They are not like human jaws at all but instead resemble a sharp pair of pinchers. Some kinds of Nereis use these jaws to tear off shreds of plants. Others grab animal prey with them. And some merely munch on dead bits of food in the mud.

A nereid has four eyes. However, these cannot form an image the way our eyes do. They are sensitive only to differences in the brightness of light. In searching for food, Nereis relies on its senses of smell and touch. It has a pair of chemically sensitive "nuchal organs." These are pits containing nerve cells which respond to food aromas. They can be pushed inside-out to form little knobs. The nuchal organs are very important in feeding. If they are destroyed, the worm no longer eats. Nereis also has head appendages of several kinds which are sensitive to touch. It has a pair of short tentacles, a pair of fat sensory palps, and four pairs of long, thin cirri with which it can feel its way about.

The British use a kind of Nereis as bait for fishing. They call it the ragworm because its body looks like a rag with limp shreds when removed from the water. The "shreds" are actually paddles, which are very useful to polychaetes. Each segment has a pair of these paddles, called parapods.

The sandworm Nereis (or Neanthes, according to some biologists) has sensitive tentacles and palps on its head (at right), and many segments, each with its pair of parapods.

DR. WILLIAM STEPHENS

There is one parapod on the right side and one on the left. Each parapod has two lobes, one on top and one below. The lobes are supported inside by stiff spines. Each lobe also has a bundle of bristles or hairs on the outside called setae, which fan out from special pockets. The name "polychaete" means "many bristles," and polychaetes certainly have lots of them. The paddle-like parapods and fan-shaped bunches of setae help a nereid grab the sides of its burrow, crawl along the bottom, and swim. The parapods and setae of other polychaetes have shapes that suit their different ways of life, as we will see.

### Inside the Worm

While the outside of a nereid is obviously divided into individual segments, inside there are three body systems which unite the animal. The digestive system is quite simple. The mouth leads to the pharynx followed by the esophagus, intestine, and rectum. Nereis itself lacks a stomach, but many other polychaetes have one.

The nervous system is quite simple, too. In the head is a lump of ganglia, or knots of nerves—sometimes called a brain. This sends nerves to the sense organs on the head. Two nerves from the brain encircle the pharynx and unite on the lower side of the worm. There they form a nerve cord which extends the whole length of the worm. The nerve cord gives off several branches in each segment. Some are motor nerves, which control muscle movement. Others are sensory nerves, which receive impulses from sense cells under the skin.

Nereis is a heartless animal, but it has well-developed blood vessels. The blood is pushed through them by contractions of the vessels themselves. One large duct runs above the digestive system. This one makes especially strong contractions which push the blood forward. Toward the front end of the worm are several vessels which go around the gut and connect the top, or dorsal, vessel with another one below the gut, the ventral vessel. Blood flows backward in the ventral vessel, and branches lead from it to the parapods. There are small tubes in the parapods which lie close to the skin. The blood sloshes around in these a bit before returning to the dorsal vessel. Here in the parapods, oxygen is taken up by the blood, and wastes, such as carbon dioxide, are released. In this way the parapods of Nereis act as primitive gills.

A nereid has organs called nephridia which remove other waste products from the body. Each nephridium has a funnel-shaped opening into the coelom. The funnel is lined with cilia which beat inward, pulling fluid from the coelom into the long, coiled tube of each nephridium. Waste materials are concentrated in the tube and released to the outside through a small pore on the underside of the worm.

The body plan of Nereis is not greatly adapted to one way of life. The worm is good at burrowing, crawling, and swimming. There are many kinds of Nereis which live in different types of places. Their generalized body plan enables them to survive in many environments. Some species are fairly small, such as one 20 centimeters (about eight inches) long that is found on shell-strewn sea bottoms from Long Island, N.Y., southward. Cooler waters with mud flats are more suitable to the very common Nereis found on both our coasts and in northern European waters as well. One of the largest and most impressive of all polychaetes is about a meter long (over three feet), a species found along our Pacific coast. These stout worms are a startling sight when they swim up out of the deep at night, their iridescent skin glowing in the light of a fisherman's lamp. Perhaps such undulating, glowing swimmers were partial inspiration for sea serpent legends.

In polychaetes which have a less flexible life style, this basic plan is modified in many ways. In burrowers, there is a tendency for different regions of the body to become specialized for different functions. More highly developed gills may be present on the head or parapods, since oxygen may be scarce in a burrow. The parapods of crawling worms may point downwards, while those of swimmers may be especially large and flat, an aid in swimming. Polychaete setae vary tremendously in size and shape. Burrowers have short setae, used to grab onto the walls of the burrow. The setae of swimmers may be flattened and paddle-like. The fireworms have long, hollow setae which contain poison. They break very easily and can cause a painful wound to the mouth of an inexperienced predator or the foot of an unsuspecting swimmer. The secret of the polychaetes' successful variety lies in the way the basic body arrangement has been changed by evolution and so suits these different ways of life.

## The Catworm

One of the strongest polychaete swimmers is the catworm. Its body segments are short, so the parapods are close together. The parapods have paddle-like lobes and long setae. Especially strong muscles can fling the catworm's body into vigorous waves while the parapods pull strongly against the water like oars. Like Nereis, the cat-worm is quite unspecialized in its way of life. The strong muscles and firm paddles which help it swim so well also make it able to burrow smoothly through the heavy beach sand.

This almost colorless, pearly worm actually spends a great deal of time buried in the sand. It dives its pointed head under the surface and anchors itself with its strong setae. Then it shoves out its strong proboscis and makes a hole into which it moves, using the same kinds of body waves that allow it to swim so well. The catworm's body has powerful ligaments which strengthen it so it does not collapse under the force of the sand. It has gills that are protected from the coarse sand because they are located between the lobes of the parapodia. Beating cilia, small hairlike projections, create a water current which passes over the gills even when the worm is buried in the sand.

### Endless Swimmers

Many polychaetes spend their entire lives floating in the sea. Some live near the surface, while others swim at great depths. Unfortunately, little is known about these specially modified creatures. Like many other swimming and floating animals, such as jellyfish, some of these worms are almost transparent. Tomopteris is a clear, fleshy worm with a flattened, wide body, perfect for floating. This strange polychaete lacks setae, except for one very long, slender front pair. But its parapodia are large and flat-tened, excellent for swimming. Tomopteris must be able to move quickly, for it can capture arrowworms, which are themselves speedy predators.

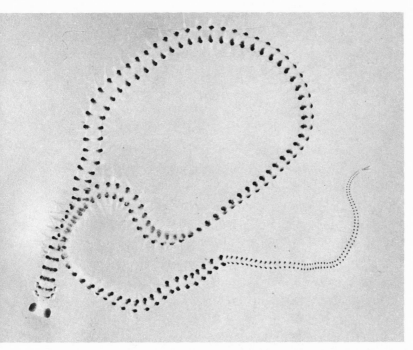

The marine worm *Torrea candida*. Its only color is in its brightly colored and well developed eyes. The black spots are glands at the bases of the parapods. Some of its setae are visible. It swims not only with the beating of its parapods but also with constant undulations of the body. Lengths vary from 12 to 17 centimeters (about five to nearly seven inches).

Some swimming polychaetes have developed very large and complex eyes, remarkably similar to those of octopuses, squids, and even vertebrates. Their eyes have structures equivalent to the cornea, iris, lens, pupil, retina, and fluid-filled parts of human eyes, as well as similar muscles and well developed optic nerves. Quite a few kinds of these worms are known from the warmer parts of the

Pacific and Atlantic oceans, but they are little studied.
They are so accustomed to life in the open sea that they
do not survive long in captivity.

The eyes of these worms dwarf the rest of the head, although they manage to have room for a proboscis, six
antennas, and sensory cirri, too. One kind which is always
found near the surface, Torrea, is about 17 centimeters
(almost seven inches) long and very slender. It swims
in the familiar annelid fashion, throwing its body into
curves and beating its parapods. Its body is clear except
for dark dots made by special glands, and its eyes are a
bright orange-red color.

A different kind of keen-eyed worm lives deeper down.
This one, called Vanadis, is short (about five centimeters,
or two inches) long and fatter, with even bigger eyes. It
too is almost colorless except for the huge, red-brown eyes
bulging up from its head. The eyes of Vanadis are most

Vanadis, another deep-water worm that has remarkably well
developed eyes. The pupils show a golden glow, a reflection
from inside the eyes.    DR. GEORGE WALD, HARVARD UNIVERSITY

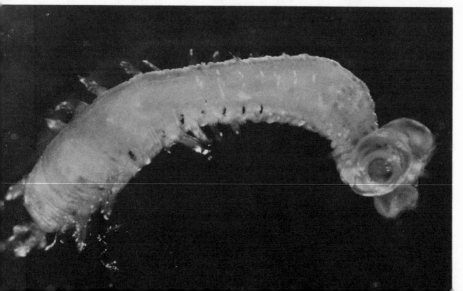

sensitive to blue light. And at the depths where it seems to live, only blue light filters through the water.

## Crawlers

Other polychaetes have given up the ability to swim in favor of more efficient crawling. An especially interesting group of crawling worms are the scale worms. The upper lobes of the parapods are no longer used in moving. Instead, every other one produces a strange stalk which carries a flattened scale. The scales, called elytra, overlap, covering the whole back of the worm. The lower lobes of the parapods are stout and cone-shaped. They are used for walking. Scale worms march along with little of the typical polychaete side-to-side motion. Their shorter, plumper but flatter bodies lack the tissue walls between segments that allow for such movement.

The elytra are quite useful. They help protect the worms when they are lodged in crevices, a favorite hiding place of scale worms. The space under the elytra provides a channel for the worm's respiratory current. Cilia pull the water along the worm's back under the scales, so even when it is jammed into a small space the worm can get fresh oxygen-rich water.

The elytra protect their owners from predators, too. Some scale worms can shed them and leave an enemy with a useless mouthful of scales. Others of these worms break in two if they are attacked. The front end crawls off to safety while the rear stays, writhing and twisting. Meanwhile light-producing organs on the elytra flash rhythmically. The predator is distracted long enough for the front end to escape. The elytra of some scale worms provide a protective covering for the developing eggs and larvas, too. It is a perfect place for them, hidden from enemies and well provided with oxygen by the respiratory current of the mother.

### The Sea Mouse

The sea mouse Aphrodite is a scale worm which hardly seems wormlike. It does not really look like a mouse, either, although the name points out the most peculiar trait of this unusual worm. The sea mouse has elytra like other scale worms. But one cannot see them, for the whole back of the worm is covered by a feltlike layer of fine setae. Along its sides are ridges of long, sharp setae as well. When first dug out of the mud, Aphrodite is a dull, muddy gray color. But if it is left in clear water for awhile, the mud washes off and the sharp setae flash with a lovely iridescent glow.

Aside from this, the sea mouse is a dull-looking creature. Its short, fat body is quite ungraceful and almost totally incapable of any side-to-side motion. Aphrodite crawls along just under the surface of the mud, stepping with its strong parapods. Only the tail end is exposed to the water. The thick layer of felt protects the respiratory channel under the scales from getting fouled by the mud. Water is pulled in under the tail end and passes up between the parapods into the space between the back and the elytra. Then the elytra are closed down, starting from the front end, pushing the water current out above the tail. This slow but sure mud-dweller represents the extreme modifications of the basic "plan" for swimming, burrowing, and crawling polychaetes.

*Four*

# Lives in Bank-Vault Homes

The variety of burrowing and tube-dwelling polychaetes is tremendous. While some of the crawling and swimming polychaetes spend their hours of rest inside burrows, other worms remain buried for their entire adult lives. We have seen how some of these bristled worms are specialized for floating and swimming and how others have become very efficient crawlers. Burrowing worms also show special adaptations for their limited lives. Their parapods are usually much smaller than those of swimmers and crawlers, since they are of little or no importance in moving. Their setae are generally shorter, too, since their only function now is to grab the sides of the burrow.

While some crawlers and swimmers have protective devices such as poisonous bristles or hardened scales, the bodies of burrowers and tube-dwellers are generally soft and unprotected. Their sense organs are reduced, too. They lack sensory palps and the long, antenna-like cirri. If they have eyes at all, they are of a much simpler type than those of the more active worms. These inactive animals are very well adapted for their special life styles, however, and many of them are extremely common in their particular environments. Great numbers of lugworms can be found on sandy tidal mud flats, and fanworms often adorn rocks and reefs with large, colorful colonies.

### Tunnel Networks

Several types of polychaetes live in networks of burrows in the sea. They line their complex galleries of interconnecting tunnels with slimy mucus and move around within them. Many such worms look very much like earthworms. Their heads are small and pointed and may completely lack eyes, palps, and antennas. Instead of being large and lobed, their parapods are often small, with reduced setae which do not interfere with burrowing.

One especially interesting gallery dweller is Glycera, the marine bloodworm. Glycera constructs several looped connecting tunnels in the muddy bottom which have many openings to the surface. The worm lies in wait for prey within a tunnel. Through its tunnels it can feel changes in water pressure caused by animals crawling over the mud. When the worm knows the path of its victim, it inches up close to the surface. As the unwary prey passes over the tunnel opening, Glycera shoots out a long, strong proboscis. Four sharp pointed jaws grab the victim and inject poison into it. Then the worm quietly pulls its dinner back down inside its home to eat in safety.

### The Life of the Lugworm

The lugworm, Arenicola, lives in a simple L-shaped burrow which has one opening at the top. The tail end of the worm sticks up toward the surface while the head faces the blind end of the burrow. Arenicola literally eats its way through the sand and mud, slowly but surely. It has a short, fat proboscis which it pushes out and in, out and in, swallowing mouthfuls as it goes. Muscular waves of the body pull water into the tail end of the burrow. The water passes the body of the worm and goes out by filtering through the sand by the worm's head. Any food particles suspended in the water become trapped by the sand in front of the worm. These, plus any plant and animal remains mixed in with the mud and sand, provide this slow but steady feeder with enough food. Every now and then, the worm backs up in its burrow and deposits a neat coil of rejected mud and sand. To a human observer, these coils around the circular burrow openings are clues to the presence of lugworms on a beach or mud flat.

The worms we have looked at so far have segments which are pretty much identical. But Arenicola is different. Its body has regions where the segments have various functions. Segments near the head have smaller parapods. The two lobes of each are separated and have short setae. These are used to grip the sides of the burrow. Nephridia for excreting wastes are found only in these body segments. Segments in the next group have parapods with setae and gills. This part of the body is where oxygen is obtained from the water and carbon dioxide is released.

The tail end of Arenicola lacks setae, gills, and nephridia. It is simply a muscular tube which contains the sand-filled rectum. This end, which is extended from the burrow to deposit food wastes, is easy for predators to nip off. The tail is very sensitive and can be jerked rapidly back into the burrow. But even if the worm is not quick enough and loses its tail, nothing vital is lost, and a new one is rapidly regrown.

## The Ultimate Tube Dweller

The strange body of the worm Chaetopterus may look bizarre and helpless outside of its burrow. But when one learns how that body functions within the flexible U-shaped tube which it secretes, it is easy to see why this worm is so common and successful. The body of Chaetopterus is divided into many sections, each specialized for a particular function. About at the middle of its body, Chaetopterus has three segments with greatly enlarged lobes of the parapods. These lobes are fused in the middle, forming three fan-shaped paddles which beat powerfully about once a second. They create a strong, continuous water current within the tube. Near the head, this worm has another enlarged pair of parapods. These, however, are long and winglike. They are held outstretched, pressed against against the sides of the tube, with their tips meeting at the top.

These parapods secrete a sheet of mucus which forms a bag. The bag is dragged backward by the water current until it touches a special cup-shaped structure farther back on the body. The cup holds the end of the bag. Water entering the burrow filters through the bag, and food particles as small as bacteria are trapped in the mucus. As the bag is loaded with food, the cup gathers it together into a little ball. When the ball reaches a certain size, it is cut off from the mucus bag. The cup then places the ball in a groove lined with cilia. These cilia beat toward

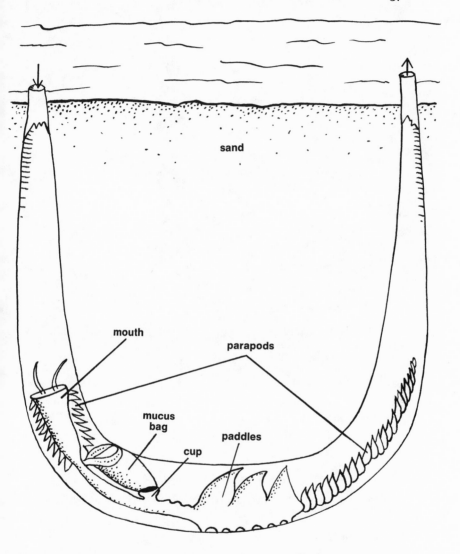

Chaetopterus lives in a U-shaped parchment-like tube that it constructs under the sand. It is one of the oddest-looking worms there is, but its curious anatomy is quite efficient for its kind of life. Arrows show entrance and exit for sea water.

The fanworm *Eudistylia polymorpha* emphatically contradicts the notion that "all worms look nasty."

the front of the worm and carry the food ball up to the mouth, where it is swallowed. Each ball carries with it just about every edible particle and microorganism present in one cup of sea water. Cilia around the mouth can detect any large objects which are pumped into the tube. To avoid them, the worm merely raises the wing-shaped parapods and lets the objects pass down the sides of the tube outside of the mucus bag.

## The Beautiful Fanworms

Fanworms, also called feather duster worms, are the flowers of the sea. Their graceful, feathery crowns come in an assortment of bright colors and wave gently in the water, like flowers in the wind. Watching them, so quietly outstretched, one finds it hard to believe that these colorful "blossoms" are parts of animals. But just try to touch one; it will disappear like magic. Where did the flower go? Looking more closely, one would see a small tube that looks much too small to hold that graceful fan. By watching patiently, one could see the crown slowly emerge from the tube and spread out again.

While spread and relaxed, the crown of the fanworm looks inactive. What good is it to its owner? In reality, it is always at work collecting small food particles from the water. The crown consists of individual branches called radioles. Each radiole in turn has projections called pinnules, which give the radioles their feathery appearance. The radioles and pinnules are covered with cilia. The beating of these cilia pulls water through the crown from outside to inside. Any food particles in the water are trapped by mucus on the pinnules and carried by cilia to the top surface of the radiole. Here there is a groove which carries food particles to the mouth. But the pinnules have no way of telling which particles are good for food and which are not. Generally, the smallest particles are most likely to be food, and the radioles have a way of separating out the food from the other materials.

This system is best developed in the fanworms called sabellids. Sabellids can get quite large, with colorful

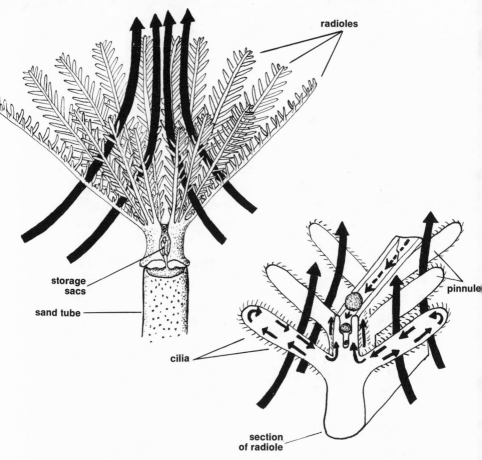

At top, the fanworm Sabella is seen with its crown of radioles extended. The lower picture is a magnified view of three pairs of pinnules on one radiole. Thick arrows show the direction of water that is pulled in by the cilia on the pinnules and radioles. Small arrows show the path of food and other particles as they are carried by the cilia along one pinnule and to the top of the radiole. Dashed arrows show the path of particles down the groove toward the mouth. Notice that oversize particles are carried on top of the groove; they will be discarded. Middle-sized pieces, which will be stored in sacs and used for tube-building, are carried by ridges inside the groove. The smallest particles drop to the bottom of the groove and are swallowed.

crowns as big as eight centimeters (about three inches) across. These worms build tubes of sand and other particles mixed with mucus. The radioles of sabellids have grooves with three ridges. The outer edges of the groove form the largest groove, which is on top. Large particles are caught on this groove and are carried down to the base of the crown and discarded. Middle-sized particles are caught by the ridges inside the groove. These are carried to sacs near the mouth where they are stored for future use in tube-building. Finally, the smallest particles are carried in the bottom of the groove to the mouth, where they are swallowed.

Sabellids are very efficient tube-builders. The sacs which store the sand particles also secrete mucus. In order to add to its tube, the worm mixes the sand and mucus and lays it out in a string along the top of the tube. This forms a collar, for while the new building material is being poured out, the worm rotates slowly in the tube. The rings of the collar mold the coil of sand to the top of the tube. A pair of glands on each segment secrete mucus which lines the entire tube, making it comfortably smooth inside.

The other kind of fanworm is called a serpulid. These worms are generally smaller than sabellids, with crowns no more than one and a half centimeters (about half an inch) across. Some are minute and live attached to seaweed or snail shells. Serpulids live in hard tubes of calcium carbonate which the worms themselves secrete. The sorting mechanism of their radioles does not separate out sand grains and store them, since these worms do not need them for tube-building. Serpulids also differ from sabellids in another interesting way. One radiole of a serpulid is not

The bright-colored sabellid fanworm Spirographis.

used in gathering food. It has a flattened top instead which fits perfectly into the opening of the tube. When a serpulid pulls inside, this modified radiole is used to plug up the hole at the top of the tube like a cork in a wine bottle, effectively protecting the soft body of the worm inside.

## Coiling Tentacles

Perhaps you have seen terebellid worms in a film about the sea. They are strange creatures indeed, with masses of coiling, squirming tentacles reaching this way and that over the sand. In films, these worms are usually shown out of their burrows where they don't belong. The tentacles are then used to pull the animals around until it can find a safe place under a rock or until it can make a new burrow. But its real home is in the sand or mud. From its burrow the terebellid spreads out its tentacles in all directions. These have muscles in them and can be stretched out, coiled up, or shortened. There are cilia on the underside which help the tentacles creep over the sand.

Food particles are trapped in mucus and are carried in a groove down to the mouth. The food collects at the base of the tentacle. One at a time, the tentacles are wiped over the lip of the mouth. Terebellids can also gather larger food particles by wrapping the tentacle about them and carrying them to the mouth. These worms have tangled masses of gills which are used in respiration. The gills are blood-red, because terebellids, like vertebrates, use the red pigment hemoglobin to carry oxygen to their tissues.

## Living With Other Animals

Some polychaetes live protected lives by taking advantage of the shelter offered by other animals. Scaleworms especially often live in the tubes of other worms or among the spines of sea stars. One kind lives in the mantle cavity of the common keyhole limpet on the Pacific coast. The setae of such worms are often modified into tiny hooks which help them hang onto their hosts. In at least some cases, the worms are attracted to their hosts by chemicals.

"Guests" other than worms may join tube dwellers, too. A kind of crab shares the tube with Chaetopterus. The crabs enter when young and stay in the tube the rest of their lives. Almost always, one male and one female crab live with one worm. When the crabs mate, the larvas are swept out of the tube with the worm's water current and —sometimes, at least—find hosts of their own.

Few polychaetes are actual parasites. One very small worm is so modified for life on its lobster host that scientists assume it is a parasite. This worm is only a quarter millimeter to one and a half millimeters long. It has a fat head and only five body segments, with a single large pair of parapods at the rear end. Using these "legs," the little worm walks over the gills of the lobster. When a female lobster is carrying eggs, the parasite moves to the egg mass and probably feeds there. These worms cause no obvious harm to their hosts, however, even when more than 600 share one host.

The most peculiar polychaete parasites are the myzostomes. These worms are so specialized as parasites that many scientists consider them to be a separate family from polychaetes. But their obvious segmentation, hook-like setae, and typical polychaete larvas show them to be at least closely related to polychaetes.

Myzostomes are always found living on or inside the bodies of echinoderms. Some live with sea stars or brittle stars, but most of them infest sea lilies. The myzostome's body is rounded and flattened. It is only a few millimeters long, with five pairs of parapods, each with one hooked seta. The parapods are located on the bottom side. Some myzostomes are not true parasites but merely use their echinoderm hosts as food-gatherers from which to steal nourishment. But others live in the body wall, gut, or gonads and feed on the unfortunate host's tissues.

*Five*

# How Polychaetes Reproduce

Many polychaetes can reproduce by budding, or simply growing a new individual from the side or end of the parent worm. Some budders form chains of several growing worms, while others may even produce branching chains of new individuals. But the most common form of reproduction among these worms is the mating of separate males and females.

Most polychaetes have a very simple reproductive system. Males have no definite testes and females have no true ovaries. The sperms or eggs are simply budded off from the lining of the coelom. The eggs or sperms, called gametes, then mature while free in the coelom. They are nourished by the coelomic fluid and by other cells budded off with them. Usually the gametes develop in only some segments of the worm. When it is ready to spawn, a polychaete is crammed with gametes. Most polychaetes have ducts which carry the eggs or sperms to the outside. But some simply burst open, dying as they release a cloud of gametes into the water.

The structure of the polychaete reproductive system is simple, but the ways of reproducing in these worms are extremely varied. Some mate and have internal fertiliza-

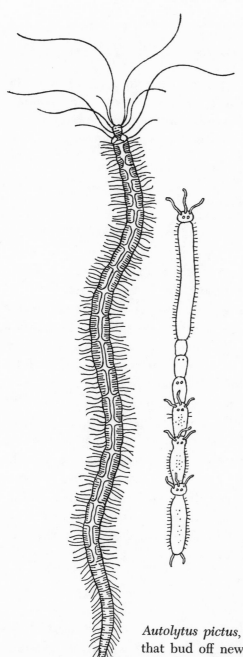

*Autolytus pictus,* one of many worms that bud off new individuals that remain attached a while.

tion. Others release gametes from their tubes, and the embryos develop nearby. Many tube-dwellers incubate their eggs. But the most common type of reproduction in many polychaete families is the fascinating method called epitoky.

### "Magical" Transformation

Burrowing polychaetes live very solitary lives, yet they must get together to reproduce. Their eggs and larvas often develop among the plankton—the drifting organisms of the seas—but how do they get there? As the reproductive season approaches, changes occur in the worms. Their bodies, which are best adapted to crawling and burrowing, change in structure in such a way as makes them more efficient swimmers. Before, the worms were repelled by light. Now they are attracted to it. The worms leave their homes on the bottom and rise to the surface, swarming with others of their kind and releasing gametes into the water. Different kinds of worms swarm at different times. This swarming behavior ensures that the male and female worms can find each other. It guarantees that the worms will pair only with other worms of their own kind. Since there are endless variations on this interesting way of reproducing, we will look at a couple of examples in more detail.

### Nereis and Its Relatives

When the reproductive forms of Nereis were first discovered, scientists thought they were a new kind of worm. They named them Heteronereis ("other Nereis"). Later they discovered their error. But even today, the word "heteronereis" is used to describe the swarming form, or epitoke, of any worm in the Nereis family.

Many changes occur in the bodies of Nereis worms as they approach reproduction. Even though Nereis can swim quite well, its body is not adapted for a life of constant swimming in the upper waters. Its eyes grow larger, and branches of the parapods enlarge with sensory cells which help the male and female worms find each other. Other than these changes, the front parts of the worms do not change. But the segments toward the rear, which contain the ripening gametes, have more extensive changes. The color may alter. New fleshy lobes grow on the parapods. These have many blood vessels and probably help in the increased respiration which occurs with ripening gametes and more activity. The bunches of fine, thin setae of the bottom dweller are replaced by fans of strong, paddle-shaped setae which act like oars in swimming.

Changes are occurring inside the animal as well. The muscular system is completely rearranged. Some muscles become smaller and others grow larger. The type of muscle fiber changes, too. The result is a worm with a type and arrangement of muscles well suited to the active, constantly swimming life which it will soon live.

### The Palolo Worm

When Nereis spawns, the whole worm swims to the surface. But in many polychaetes, only part of the worm participates in swarming. A reproductive section called a stolon is produced at the rear of the worm, while the front end remains unchanged. When swarming time comes, the stolon, crammed with gametes and adapted just for swarming, breaks off and wriggles to the surface. The rest of the

worm remains below, continuing its quiet, solitary ex-
istence.

The most famous worm of this type is the Pacific palolo
worm. Famous though it is for its spectacular and predictable swarming behavior, this worm has hardly been
studied scientifically. Only in the 1960s did scientists
finally take the trouble to examine these worms carefully.
The palolo worms live in the coral reefs of Pacific islands.
Any one worm may have as many as a thousand segments.
The body of the palolo is divided into three regions. The
front part includes the head and a series of segments lacking gills. This is followed by the middle region, with
gills attached to the parapods. The third section is the
reproductive part. It has segments which are longer and

The rear end of the palolo worm turns (at arrow) into reproductive units, making a section called the stolon. This rises to
the surface of the sea at swarming time. The reproductive
units are not all shown; they can number 800 or more.

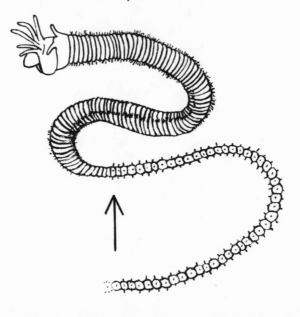

A leaf of palolo worms on sale in Samoa. Eaten raw, the worms are often considered not very tasty, but when cooked with butter they have a reputation as a real delicacy.

narrower than those in front. Each segment has a light-sensitive eyespot on the underside. The body wall of the reproductive segments is thinner than in the other segments and their muscles seem to be more active.

The stolons of the palolos break off, rise to the surface and swarm at predictable times. These times are related to the phases of the moon. There are two swarming times, one in October and one in November. One time many worms swarm, while the other time fewer do. The swarming occurs about a week after the full moon. During the big swarm, thousands and thousands of worms rise to the surface, writhing and twisting together until the segments burst, releasing countless gametes.

For hundreds of years, the natives of Samoa have celebrated the coming of the palolo, for these people consider the gamete-packed worms a special delicacy. They gather on shore the nights when the swarming is expected. Dancing and singing highlight the waiting period, while by the reef men keep watch for the coming of the worms. When the wriggling creatures appear on the surface, a happy call from the waiting men brings the rest of the people splashing excitedly into the water with nets, buckets, and flashlights.

They have only an hour or two to catch whatever worms they can collect before the animals "melt" and spawn. The collected worms are drained of sea water to keep them from "melting." The Samoans cheerfully eat handfuls of raw worms, but also enjoy them wrapped in leaves and broiled until they turn a dark green. The palolo can also be preserved for use throughout the year by cooking until the worms are dry. This dried palolo is kept in the chief's house and eaten only on special ceremonial occasions. In years when there is preserved palolo left over, it is divided among the other families and can be eaten at any time. This dried food is mixed with coconut cream and recooked before being eaten.

### Other Polychaetes

Many polychaetes do not develop epitokes or stolons and swarm. Some just leave their eggs on the sand or mud, but others brood them. Many tube-dwellers brood their eggs within their protected homes, while some scale worms carry their eggs around hidden under the elytra. The eggs of some kinds simply stick to the body of the female until they hatch. One kind of Arenicola lays her eggs in a long jelly-like mass. One end is anchored in the sand, while the other end floats loosely with the waves. After the young hatch, they remain inside their protected home, eating the mucus of which it is made. They remain for quite a while, sometimes until their bodies have sixteen segments. Then they leave and swim for a while before settling down and burrowing like their parents.

### How Polychaetes Develop

Many kinds of marine animals release their eggs and sperms into the sea water and leave them there, as so many polychaetes do. Snails, fishes, sea stars, and sea urchins of various kinds reproduce this way. There are several advantages to this system. Adult animals need not waste time and energy caring for their young. The ocean currents help spread out the larvas so that they do not end up too densely concentrated in one area. And the young animals live in a different place and feed on different kinds of food than the adults do, so they do not compete with them. One disadvantage is that many eggs and young get eaten by other hungry animals. But since an egg of these creatures develops rapidly into a feeding larva, females can produce many small eggs without having to use a lot of energy making yolk to nourish the developing young.

Shortly after being fertilized, the polychaete egg divides for the first time, producing two cells. Then each of these divides to make four cells, and so forth. Within a day or two, there are thousands of tiny cells. The shape of the larva begins to become visible. The polychaete larva is a simple but very efficient type called a trochophore. Many snails, too, have trochophore larvas. This is one reason scientists feel that annelid worms and snails are closely related animals, despite their many differences.

The trochophore is top-shaped and has two bands of cilia around its body with which it swims. At the upper end of the trochophore is a tuft of cilia, while there is a small circle of cilia around the bottom end. Between the center bands of cilia is the mouth. There is an expanded stomach and an anus at the bottom end of the larva. The trochophore feeds on minute plants which live in the plankton. Within a few days or weeks, the adult worm begins to form in the trochophore. The area between the mouth and anus gets longer and longer. Gradually, parapods develop which later grow setae. Meanwhile the embryo above the mouth region forms the front end of the worm, including the brain. While all this is happening, the larva of most polychaetes stays in the plankton. Some even develop exaggerated setae that help them float and give them protection against enemies.

### Finding a Home

Once the polychaete larva is ready to settle down, it must find the right kind of home. It swims here and there, testing out the bottom to find just the right sort of place. Each kind of larva responds differently to what it finds, for each kind of worm lives in a different environment. The larvas of sand-dwelling worms respond to the size and shape of the sand grains, or to the sizes of the spaces between grains. Bacteria which live on the surface of sand grains may also produce chemicals which attract certain polychaete larvas, while dead organic matter repels them. Some larvas are especially particular about where they settle. One kind of minute fanworm will settle only on a particular kind of algae, while a close relative chooses a different alga. These choices are probably due to chemicals produced by the plants. Some larvas will accept a second-best place to settle if that is all that is available. But others will swim until they die rather than give up the search.

# *Six*

# Earthworms and Their Relatives

Earthworms are certainly the most familiar worms to just about everyone. Earthworms of one sort of another are found living in soil everywhere on earth except in deserts. They have attracted the attention of great scientists like Charles Darwin and of fishermen and gardeners everywhere. And recently, the possibility of using earthworms as food for humans is being explored. The earthworm may be "lowly," but it is one of the most helpful of nature's creatures. It is also one of the most interesting.

Earthworms belong to a group called the oligochaetes. While the name "polychaete" means "many setae," the word "oligochaete" means "few setae." Oligochaetes have no parapods. Their setae are usually short. Muscles at the bases of the setae are used to push them out or pull them in. Oligochaetes have a poorly developed head. They lack eyes, palps, and cirri. A very few kinds have a long tentacle on the front end of the body. Many oligochaetes, especially some common fresh-water kinds, are less than half a millimeter long, while Australia is host to giant earthworms more than three meters (ten feet or so) in length.

Adult oligochaetes have a distinctive swelling somewhere along the front half of the body. It is quite obvious on the earthworm. This "clitellum" is very important in oligochaete reproduction, as we will see later. The segments of the body are obvious in oligochaetes, both large and small. Many of the small fresh-water kinds are transparent and their internal structure is easy to see. The circulatory system is like that of polychaetes. The dorsal vessel contracts, pushing the blood forward. Many oligochaetes also have vessels, which also contract, connecting the dorsal and ventral vessels. The common earthworm has five of these "hearts" near the front end. Like polychaetes, most oligochaetes have one pair of nephridia in each segment. Their nervous system, too, is similar.

## The World of the Earthworm

Have you ever wondered what it would be like to be an earthworm? Your body could stretch out at great length or be bunched up short and fat. You could broaden the rear of your body and anchor yourself by your setae while stretching out your front end to explore. You would not be able to see, but you could sense the presence of light with receptors scattered all over your body. While you would have no nose for smelling, you could sense chemicals with any part of your body. Try to imagine yourself crawling through the soil, nosing your way around rocks and eating dirt. What a strange life, compared to living out in the sun, experiencing the world in the many ways we humans can.

But earthworms have everything they need for the kind of life they live. They are well adapted for a burrowing life. It is important for them to sense the chemistry of the soil around them and to stay in the damp earth. Their thin skin has to remain moist, since it is used to pick up oxygen and release carbon dioxide. And it must be protected against poisonous chemicals which could quickly damage it.

The digestive system is well suited to feeding on the dead plants and animals which the worm encounters as it eats its way along. The earthworm has a strong pumping pharynx. It places its mouth up against a bit of food and the pharynx pumps it in, where it is mixed with moist saliva. After its esophagus, the earthworm has a crop for storing extra food. It also has a gizzard for grinding

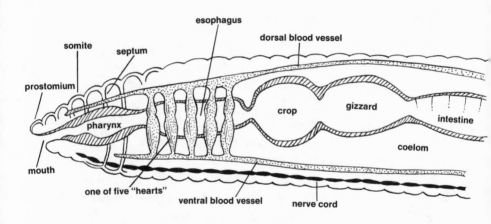

Some internal organs of the earthworm, shown in simplified form. There are ten so-called hearts—five on each side of the gastric system. The pharynx lubricates food and pushes it on to the esophagus, which neutralizes any acids that are present in it. The crop stores food temporarily, then it goes to the gizzard, which grinds it up with the help of sand grains.

food. The gizzard has a hard lining and strong muscles which crush large particles. The esophagus also has specialized glands for removing excess calcium which the earthworm has taken in with its food.

By the time the food has reached the intestine it is ready to be digested. The intestine produces chemicals called enzymes which break down the food. Earthworms are unusual in having an enzyme which can break down the tough cellulose found in plants. The digested food is then absorbed by the walls of the intestine into blood channels which lie underneath it. The intestine has a fold which hangs down from the top and runs the whole length of the intestine. This fold provides greater surface area for the absorption of nutrients.

### Earthworms as Soil-Turners

Since we rarely see earthworms, we do not realize that they swarm under our feet by the thousands. Nor are we aware of how very important they are to us. In most soils, one-half the weight of animal life is earthworm (much of the remainder is nematode worms). An average acre of ground contains a half ton of earthworms, with up to 12 tons in an acre of rich soil. Tunneling and feeding under one square meter of an average pasture are about 800 large earthworms and perhaps 8000 smaller ones. Earthworms bring from two to a hundred tons of material to the surface over one acre of land in a year, depending on the soil and the climate.

Such large numbers of animals have important effects on the soil. Earthworms do a great deal to mix it up. They pull leaves from the surface down into their burrows. They

move deeper soil to the top when they deposit their castings. Because of this activity, stones and other objects
lying on the ground are gradually buried. Earthworms
pull soil out from underneath and deposit it on top. Over
a few years, several inches of soil may be deposited on
top of things which used to be lying on the surface. Earthworms are a major cause of the burying of ancient cities
and monuments.

All this activity helps improve the soil. When earthworms pull leaves into their burrows, shred them, digest
them, and deposit the remains, they are making valuable
nutrients available to plants. Their tunnels allow air and
water to penetrate soil easily, and they provide easy channels for root growth. The castings and decaying leaves
lining the burrows nourish the roots.

## Earthworm Farming

Worm farming is said to be the fastest-growing agri-
cultural industry in our country. Although only a few
people try to make a living from earthworms, probably
50,000 persons grow worms part-time. Sheltered by garages
or old shacks, in discarded bathtubs or specially built
troughs, earthworms are earning money for many Ameri-
cans. Most growers raise a special hybrid type of worm.
While it is smaller than the common "night crawler"
which lives in lawns and gardens, it is a superworm in
terms of its food consumption and rate of reproduction. In
one year these worms can reclaim an amount of soil that
would take ten years to revive using regular fertilizer
treatment. They can double their numbers in 60 to 90 days.
They eat their own weight in food every day and will

consume anything organic, including old rags and paper. (Earthworms are not poisoned by the lead in printer's ink, for special cells in their bodies store it in a nonpoisonous form.) But the chief food for the worms is manure. As they grow and reproduce, making more worms to sell, the worms also turn the manure into rich, dark fertilizer which can be sold for garden use.

The worms themselves are bought for many different uses. Gardeners buy them to improve their soil. Fish hatcheries, game bird farms, and pet stores buy them for animal food. Scientific researchers need them for experiments, and fishermen use them as bait. Experiments are under way using earthworms to dispose of the vast quantities of garbage and sewage sludge produced by cities and turn it into safe, useful fertilizer.

## Earthworm Cookies

The latest new use for earthworms is as human food. Yuck, one might say. The idea of eating worms is certainly not appetizing at first. But they are 75 per cent protein, low in fat, and could provide a fine source of nourishment for people, especially if the world food shortage predicted by many scientists becomes a reality. Home economists around the country have been developing worm recipes, and worm farmers have been sponsoring recipe contests with prizes for the winners.

It takes a bit of work to get worms "recipe-ready." They must be fed corn meal for two days and then placed in several changes of boiling water. But once prepared, their possible uses seem limitless. They can be chopped and added to soups or casseroles. They can provide the main meat dish for a meal, as in "Earthworm Patties Supreme," which was entered in the North American Bait Farms 1976 recipe contest. Or they can be dried in the oven, crumbled, and used in breads, muffins, or cookies as a substitute for raisins or nuts, as in another recipe from the same contest called "Oatmeal Earthworm Cookies." The only problem with using earthworms as human food is that, even after reading the recipes, most people still say "yuck."

## Making More Earthworms

One of the strangest things about earthworms is their way of reproducing. It seems very complicated to us humans, but it works well for the worms. Each earthworm has both male and female reproductive systems, so each worm makes eggs and sperms. When mating time comes, two worms meet on the soil surface at night. They lie next to each other with their front ends pointing in opposite directions. The clitellum of each worm makes a lot of mucus. Each worm becomes encased in a tube of this slime. They are attached to each other so that the clitellum of each worm lies against a part of the other worm which has a special sac for storing sperms.

Setae in the clitellum of each worm pierce the body of the other worm, helping to hold the two firmly locked. Around the clitellum, the slime tube encircles both worms, which also keeps them together. Each worm releases sperms from a pore a few segments in front of its clitellum. The sperms move backward in a groove along the body. When they reach the clitellum, the sperms pass across to the other worm through a gap in the slime tube. They enter a small opening which leads to the sperm storage sac. This whole process takes two or three hours. When

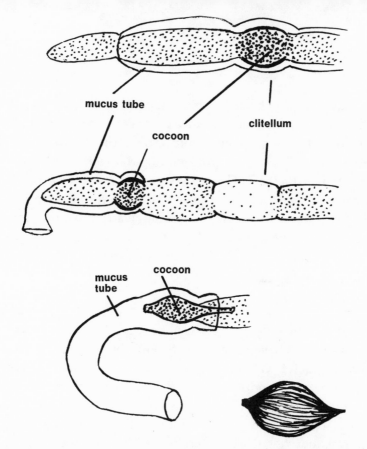

The egg cocoon of an earthworm is formed by the clitellum and gradually slips off the front of the animal in a tube of mucus. The ends of the lemon-shaped cocoon (shown enlarged in last picture) close and the cocoon, now dark in color, is deposited in damp earth.

each worm has finally received sperms from the other, the two worms separate.

A few days later the worms are ready to lay their eggs. The clitellum of each animal secretes a protective cocoon for the eggs. The outside layer of the cocoon is tough and resists drying. The next layer is made up of albumin that nourishes the developing eggs. The newly made cocoon slips forward along the worm's body. As it moves, eggs are released into it from the female pores. Farther forward, the sperms which were received from·the other worm are deposited among the eggs and fertilize them. The cocoon then slips off over the front end of the worm and is deposited in the soil. The ends seal shut, making a moist, protected place for the young worms to develop. In two or three weeks the young worms hatch and crawl out from the ends of the cocoon.

### Earthworm Relatives

You might be surprised to learn that ponds and lakes often swarm with many kinds of oligochaetes. The bottoms of lakes may have as many as 8000 Tubifex worms in one square meter (about ten square feet) of mud. Other worms, some microscopic in size, crawl around among the water plants. Although little is known about the lives of these creatures as compared to earthworms, interest in fresh-water oligochaetes has perked up in recent years.

A major reason for this is the concern with pollution. Tubifex and some of its relatives can live in water with very little oxygen. They can even survive for short periods of time with no oxygen at all. When water becomes polluted with sewage or fertilizers, the amount of oxygen in the water is greatly reduced. Most water animals die, but the population of Tubifix soars. Now it has no competition for food and space. Its predators have disappeared, too, so it can multiply very rapidly. Different kinds of Tubifex and its relatives need different amounts of oxygen to survive. So, by studying the numbers of worms of different kinds which are present, scientists can learn a great deal about the severity of pollution. By sampling the worms at different times and noting changes in those populations, they can keep track of changes in the degree of pollution as well.

How does Tubifex survive so well where other animals die? It is not a very impressive creature. A giant of its kind would be an unimpressive 20 centimeters (about eight inches) long. Most Tubifexes are about three or four centimeters (about one and a half inches) long. These worms have short setae and little in the way of a head. They are very wormy-looking worms. They live buried in the mud head first, with their tails sticking up out of their tubes.

The worms are also called "bloodworms" because of their deep red color. That color is due to the hemoglobin in their blood, one secret of their success. This Tubifex hemoglobin is very good at collecting oxygen from the water. When there is plenty of oxygen, these worms have only a bit of tail sticking up from the mud. But if little oxygen is present, they raise more of the body. They wave their tails back and forth. This helps circulate the water and exposes them to as much oxygen as possible. They pull water into their intestines and thus bring the oxygen close to their body tissues.

Tubifex worms are very important in the ecology of the lakes and streams where they live. When they occur in large numbers, the water circulation caused by the combined action of thousands of worm tails waving about at the same time can be significant. Good circulation of the water helps keep oxygen and other chemicals evenly distributed in the water. It also can affect the temperature of water at different levels if the water is shallow. Tubifex feeds by swallowing mud containing dead plant and animal matter and decay bacteria. Thus they act as earthworms do on land, digesting and redistributing organic matter in the mud.

## A Minute Predator

The aquatic worm Chaetogaster lives an entirely different life from its relative Tubifex. While Tubifex is some 20 centimeters in length, a Chaetogaster is small, less than one millimeter long. The worm is very common, and is likely to show up in any sample of pond plants one might collect. Under the microscope, Chaetogaster is fun to study as it moves slowly through the plants. All of its internal organs can be seen through its transparent skin.

This worm is easy to recognize because its rather long setae occur in bunches along its body. The setae are used to push and pull the worm through the water plants. Like many other microscopic animals, Chaetogaster has a simplified body plan. Although its segments are easy to see, there are no walls between them. Because it is small, it needs a less complicated circulatory system. Even though it is a predator, Chaetogaster has a simple tubular mouth cavity. It feeds like a vacuum cleaner, sucking small worms and other water creatures into its mouth. It is often found living on snail shells. Snails have a quiet, protected pocket under their shells called the mantle cavity. Microscopic animals often live in the mantle cavity, and Chaetogaster takes advantage of this by poking its mouth in and sucking out these helpless little creatures.

## Other Water Worms

There are many other kinds of fresh-water worms. One, called Ripistes, uses its long setae for feeding. Ripistes lives in a tube and sticks out the front part of its body.

It waves its three long bundles of setae in the water, spreading them out like fans. Bits of food stick to them. Every now and then the worm cleans off the food particles with its mouth and swallows them.

The small oligochaete Dero is often found in muddy bottoms. Its rear end is modified in a rather startling fashion for collecting oxygen from the water. Dero has small lobes around its anus, or outlet of its intestine. They suggest somewhat the petals of a flower. There are many cilia on the outside of the lobes and many blood vessels on the inside. The cilia beat strongly, circulating water across the lobes and into the anus as well. This helps bring oxygen close to the hemoglobin-rich blood which flows in the vessels. When little oxygen is present, the lobes become longer, which aids in gathering the scarce oxygen.

The African swamp worm solves the problem of getting oxygen in a completely different way. It lives in water-logged muck where there is very little oxygen available. Usually the swamps where the swamp worm lives lack a water layer at the surface. The worm lives in a burrow headfirst, like so many other water worms. It sticks its tail out of its soggy burrow into the air. The tail can be flattened with a slight curve, forming a temporary "lung." The surface of this special grooved tail is richly supplied with blood vessels. Often the worm folds its tail over slightly to form a funnel. It moves down so that its hind end is about at ground level. It can stay in this protected position for more than ten minutes but will retreat down into its tube in a flash at the slightest vibration. If its burrow is covered with water, the swamp worm can reach above the surface with its tail and trap a bubble of air which it then holds in its tube.

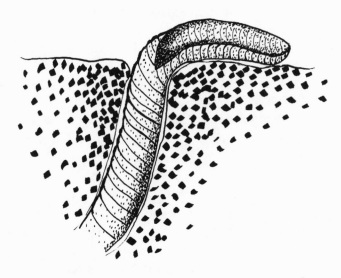

At top, the swamp worm partly flattens its specially shaped tail
into a curved form that acts more efficiently as a "lung." Below,
it protects itself from possible predators by pulling down into
its tube, but still keeps the funnel-like hind end open at the
surface where it collects oxygen.

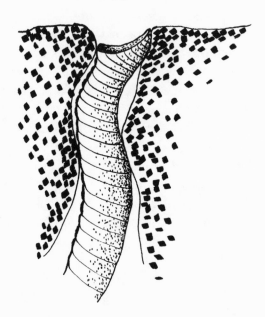

## Growing New Worms

Like flatworms and polychaetes, many oligochaetes can regenerate lost parts. If an earthworm is cut, both pieces can eventually regrow lost parts. However, this depends somewhat on where the cut is made. For example, a worm will regenerate a new head end if no more than nine segments at the head end are removed. As with flatworms, the ability to regenerate varies from one kind of oligochaete to another.

Many fresh-water oligochaetes reproduce by budding. Chaetogaster may bud off new worms, and some worms may grow in chains of individuals. There may be as many as eight small worms joined together.

Other oligochaetes can reproduce in a hurry by "going to pieces." One kind breaks up into six to eight bits, each of which forms a new worm. This process can be so fast that 15,000 worms can result in just two months from one original worm. Asexual reproduction works so well that some oligochaetes never or rarely reproduce sexually. One kind was raised for 150 generations over three years with no sign of sexual individuals. Other worms reproduce by budding most of the year and mate only during the fall.

## Seven

# Bloodsuckers and Acrobats:
# The Leeches

We have seen that polychaetes are successful with many different life styles, while oligochaetes are burrowers or fresh-water crawlers. Leeches are still more restricted in their life style and in their body form as well. For example, all leeches have 33 body segments, whether they are large or small, whether they live on land or in the water. Leeches are altogether fascinating creatures, well adapted for lives as predators or bloodsucking parasites.

For several reasons, leeches are thought to be closely related to oligochaetes. Like oligochaetes, leeches are both male and female. They, too, have a clitellum which makes a protective cocoon for the eggs. And the patterns of embryonic development for the two groups are similar.

There are major differences as well. Oligochaetes have setae. Their body segments are usually separated by septa. But leeches lack any sort of bristles and have no walls between their segments. Leeches have two suckers, one at each end of the body. By alternately attaching the front and back suckers, they move along inchworm style. Because they can move by using their suckers and since they do not burrow, leeches do not need setae. And separate body compartments which can be moved independently are also unnecessary for them. The leech has many strong body muscles to help it move. Using its suckers, a leech can inch along quite rapidly. It can hang on with its rear sucker and wave the front of its body this way and that to explore. When it is in the water, it can swim with graceful ripples of its body.

### Kinds of Leeches

There are more than 500 kinds of leeches. The smallest are only one centimeter (about ⅖ inch) long, while the giant leech of the Amazon reaches 30 centimeters (about one foot) in length. Most leeches, however, range from two to five centimeters (⅘ to two inches) long. The

Some leeches are attractive not only because of a graceful form but also because they have patterned backs.

EDWARD LINDEMANN

famous medicinal leech which was once used to bleed people as part of medical treatment for certain diseases is larger. It may be 20 centimeters (about eight inches) long. While the medicinal leech in a dull blackish brown in color, some leeches are brightly patterned with reds, greens, and glossy black.

From the outside, the leech body looks as if it has many segments, for there are narrow rings along the whole body. Each true segment, however, has from three to five of these outside rings. Some leeches have very obvious suckers, but others have a very small front sucker. While most leeches have a simple body form, some fish leeches look peculiar indeed. They have branching gills along the sides of their bodies. The front part of the body is thin, and markings on top of the expanded front sucker resemble a somewhat grouchy human face.

The skin of leeches is smooth and slippery. Many mucus cells help keep it moist. Because the skin must be kept damp, leeches are common on land only in areas with a lot of moisture. Land leeches have another special problem. The suckers must be kept wet or they will not grip well. In one group of especially successful land leeches,

Looking at a leech and looking through one. The picture at left shows the outer rings, which do not represent actual segments, and the front and rear suckers. The other picture shows a leech specially prepared for microscopic observation; it makes clear the blood-filled branched intestine at the rear and the similar-looking but lighter crop extending in front of it. The mouth shows at the front tip as a black spot; about two centimeters behind it is the sucking apparatus.

CAROLINA BIOLOGICAL SUPPLY COMPANY

the front nephridia open onto the front sucker. The back nephridia open onto body folds which touch the rear sucker. In this way, the waste fluids from the animal's body are used to moisten the suckers. The muscular sucker rim and frills of tissue around it help hold in the moisture.

While most leeches are bloodsuckers, some are predators. They have a sucking mouth which can pull in whole animals. Some of these live in ponds and streams. They eat snails, flatworms, and insect larvas. Another kind which always lives near the shoreline hunts earthworms. Some land leeches with weak jaws and blunt teeth are hunters, too. They feed on earthworms and slugs.

Other land leeches are thirsty bloodsuckers with sharp jaws. They lurk in tropical Asian jungles, attached to leaves by their rear suckers awaiting warm-blooded prey. These leeches are extremely sensitive to heat and can respond to the body warmth of a human from several inches away. In some areas, such as Vietnam, these leeches are so abundant that they fall down like rain when people pass under the leaves they cling to. The soldiers in the Vietnam war had to have every part of their bodies covered and their wrists and ankles firmly cuffed to avoid leech attack in the jungle. One small rip in a sleeve and the soldier would discover a cluster of busy leeches clinging to his arm when he took off his shirt.

The line between a predator and a parasite is sometimes hard to draw. While adult medicinal leeches do suck blood and could thus be considered parasites, some scientists prefer to call them predators, since they will suck blood from just about any vertebrate animal which comes along. They are not always associated with the same type of animals the way most parasites are. Also, they spend only short periods of time actually attached to their victims. Most of their lives they lie hidden under rocks, digesting their huge blood meals. Young medicinal leeches may act as genuine predators, feeding on small worms.

Other leeches appear to be parasites when young and predators when older. One kind lives in the mantle cavity of snails when young, taking occasional blood meals. This can be called parasitism. Adults of the same leech also attack snails. But they suck up all the body fluids and may end up taking in everything but the shell. The adult is clearly a predator. Some leeches could only be called parasites. One kind settles inside the nasal passages of a duck and is likely to stay there all its life.

The idea of voluntarily placing a leech on your skin and letting it pierce and suck may seem horrifying today. But in times past, leeches were considered valuable aids in the treatment of many diseases. People once believed that illness was caused by poisonous fluids or by spirits inhabiting the body. The only way they knew of removing these bad influences was to let them out of the body along with blood. Various methods were used to bleed sick people, including merely cutting open veins. But such methods sometimes resulted in infections or in great loss of blood. Leeches provided a more controlled way of removing blood.

Leeches were used in the Far East for many centuries, but became known in the West only when traders brought the knowledge of leeching to Greece. The first Westerner who wrote of the value of leeches lived in the second century before Christ. By the time of the Roman Empire, use of leeches was known to most European medical men —who used them so often that the physicians themselves were often called "leeches."

During the twenty centuries when leeching flourished, many famous people were treated for their illnesses by this technique—Julius Caesar, Napoleon, the Duke of Wellington, and Joseph Stalin, to name a few. Leeches were supposed to cure all sorts of ills. If one had indigestion, 20 or 30 leeches placed against the abdomen would supposedly solve the problem. Six leeches applied to each temple overcame a cold, while a neat circle of leeches placed around the head brought down a high fever. All that seemed necessary was to put the right number of leeches in the right places to cure almost any disorder of the human body.

During the early 1800s, leeching reached its peak of use in Western medicine, especially in Paris, the center of medical learning at the time. A French doctor named François Broussais, who had been an army surgeon under Napoleon for three years, became a professor at the Paris Medical School. Broussais had a very simple theory of medicine; every disease was due in some way to a digestive disorder, and leeches applied to the abdomen and/or head were the only acceptable remedy. Broussais has been called the "Bloodiest Physican in History," and his popular method of treatment led to an incredible demand for leeches. Sixty to eighty thousand leeches were sent to Paris each day, and the price of these humble little creatures soared dramatically. Medicinal leeches became extinct in England, and Russia passed game laws to protect them, as if they were elk or deer. During certain months it was against the law to gather leeches, and only specimens over a certain size could be exported. This "leech fever" did not affect the United States as strongly as it did Europe, but even so a million and a half leeches were used each year in our country during the 50 years following the Civil War.

In some remote parts of the world, leeches are still used for medical purposes, and even in the West they still have certain specialized uses. For example, leeches can drain black eyes very effectively and are popular even now with boxers for this purpose.

### How a Leech Feeds

Leeches such as the medicinal leech are very efficient bloodsuckers. The leech's mouth is located in the middle of its front sucker. It can hang on tightly while it feeds. It has three sharp sawlike jaws. The jaws slice rapidly through the host's skin, making a Y-shaped wound. The saliva of the leech contains an unknown chemical which

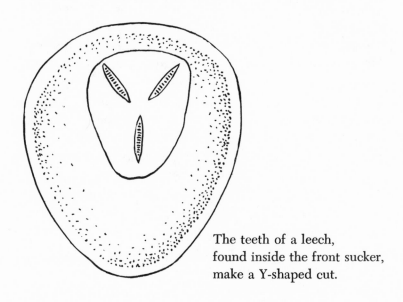

The teeth of a leech,
found inside the front sucker,
make a Y-shaped cut.

numbs the victim so it feels no pain. It also contains a different chemical, called hirudin, which prevents blood-clotting. If one pulls a leech off, the blood keeps flowing from the wound for a long time because of the hirudin that was injected.

Leeches feed infrequently. When they do find a host, they have a real banquet. The medicinal leech takes in two to five times its own weight in blood at one meal. The land leech is an even greater glutton. It can stuff in ten times its weight at one feeding. This amount of food lasts a long time. One scientist studied feeding by medicinal leeches in detail. One leech, for example, took in five times its weight in blood. Two hundred days later, it had finally digested all the food and stored up enough reserves to last for another hundred days. If a leech could feed every six months it would grow steadily, and it could survive on just one feeding a year.

How can a leech store up so much food? Its digestive system is especially adapted to infrequent huge meals. Water is rapidly removed from the blood, and this con-centrates it. The water passes from the body by the nephridia in such quantities that a land leech becomes surrounded by a pool of clear liquid as it feeds. In ten days a medicinal leech loses 40 per cent of the weight of blood taken in by removing water from it.

The gut of most leeches has many side branches that hold the blood while it is slowly digested. Leech digestion is quite peculiar. While most animals produce chemicals called enzymes which break down their food, leeches do not. Instead, their guts contain a unique kind of bacteria. These bacteria can digest the blood slowly, attacking only some blood cells at a time. They can also keep other kinds of bacteria from growing. Even after many weeks in a leech's crop, white blood cells are still intact, and red blood cells may still be there after eighteen months.

The leech depends totally on these bacteria for digestion, and they can digest only blood. If milk or egg protein are injected into a leech's gut, they are not digested. And if an antibiotic which kills bacteria is injected into a fed leech, digestion ceases. When young leeches hatch from their cocoons, they already have the special bacteria in their guts. The parent leech probably passes some of the bacteria to the cocoon when it forms a plug for it with its mouth.

### How Leeches Reproduce

Unlike many polychaetes and oligochaetes, leeches cannot regenerate or bud off new individuals. They must rely completely on sexual reproduction to make more of their kind. The testes of leeches produce sperms before the eggs are mature. When two leeches of most kinds mate, they wrap around one another, using their front suckers to hold on. Each leech attaches a packet of sperms to the clitellum of the other one. Then a strange thing happens. By a method which is not yet understood, the sperms actually penetrate the skin of the leech and wind up inside its body. They find their way to the ovaries, where they fertilize the eggs before they are laid.

The leech picks out a protected spot to place its cocoon. It uses its front sucker to apply a sticky substance. The worm then lies with its clitellum resting on the prepared place while special glands secrete the outer covering of the cocoon. Different glands produce a nutritive albumin,

and the eggs are laid in this fluid. While making the cocoon, the leech slowly turns its body, smoothing out the
cocoon lining. The front end of the animal is made long
and thin as the leech slowly backs out of the cocoon. As
they pass over the mouth, the front and back ends of the
cocoon are sealed shut by special glands in the front sucker.
Some leeches then flatten out the cocoon while it is still
soft. In a few days the cocoon becomes hard and dark
brown in color. Leeches which leave their cocoons on
land, as well as certain others, produce fancy sculptured
outer walls which probably help keep the eggs from drying out.

Some leeches actually guard their cocoons and take care
of the young after they hatch. These leeches produce a
thin-walled cocoon. Right after depositing it, the parent
leech climbs on top of it. As the embryos develop, they
grow a special knob of cells near the front of their bodies
on the underside. When the developing leeches break out
of the cocoon, these knobs become locked into certain cells
on the bottom of the parent's body, like a ball and socket
joint. Some scientists think that nourishment passes from
parent to young through this attachment stalk. But since
the embryos have plenty of yolk and the egg membranes
are still around the embryos, this is not likely.

The young keep developing while they are safely attached to the parent. Their nervous systems mature and
their suckers develop. Once this has happened, they hatch.
They still remain with their parent, getting a free ride
while hanging on with their rear suckers.

The young may remain for weeks or months. During this
whole period, the little leeches apparently continue to be
nourished by the yolk remaining in their bodies. The

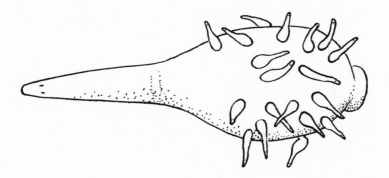

Some young leeches, such as these Clepsines, remain for a period of development on their mother's body. The actual length of an adult Clepsine is about one centimeter (roughly, half an inch).

parent leech, however, seems to provide some vital function to the young, for if they are removed too early they die. Some scientists think that the movements of the parent's body provide a fresh supply of water with oxygen to the young leeches. But such small, flat bodies should have little trouble getting oxygen on their own. Perhaps someone will study this interesting question in detail and discover for sure what vital service the parent provides to its offspring. This is only one of the unsolved mysteries of leech life. There is still much to be learned about this fascinating but little-understood animal.

# Eight

## "The Fat Innkeeper"
## and Other Curious Worms

A book about worms could go on forever—besides the animals we've already looked at, another dozen or so phyla of wormlike animals exist. Some of them are so small or so rare, however, that one would never encounter them outside a specialized zoology class. We will just take a quick look at a few of the more interesting and important kinds of other worms.

Urechis is a fat worm with little to distinguish its appearance. It has no eyes, ears, or segments. From the outside, about all that can be seen are its plump pink body and funnel-shaped proboscis. It has two setae near the front of its body and a circle of setae near the rear. Urechis lives in a U-shaped burrow and inhabits the Pacific coastline from northern California to Chile. This worm feeds much like Chaetopterus, but with important differences.

It has a ring of mucus glands around the front part of its body. When ready to feed, it moves to one entrance of its burrow and widens its body, pressing the mucus glands against the walls of the burrow. As it secretes mucus, the worm backs up, spinning a transparent mucus bag attached to the burrow at one end and to the worm at the other. Positioning itself comfortably in the burrow, Urechis then pumps water through the bag with waves of body muscle movement. When the bag is so full of food that pumping becomes difficult, the worm contracts its body, letting loose the mucus bag, which it then grabs with its extended proboscis. As it swallows the food, the worm moves back up to the burrow entrance to make another bag. If the supply of food is rich, Urechis may pump only a few minutes before consuming the bag. But if food is scarce, it may pump for as long as an hour.

As we have already seen, the burrows of animals make convenient refuges for other creatures. Since Urechis has an especially large home, it is host to several commensal animals; hence its nickname "the fat innkeeper." One visitor is a very abundant species of small fish. When the tide goes out or when enemies threaten, this fish dives for the nearest hole. Often the hole belongs to Urechis. Between tides there may be two dozen of these fish visiting one worm. Some of the fish remain between tides, feeding on large bits of food which Urechis rejects. If the fish grabs a chunk which is too large to swallow, it will carry it over to another guest of Urechis, a small pea crab. The crab will tear the food apart with its claws while the fish darts up to grab its share.

Another sometime visitor is a clam, only about one and a half centimeters (about a half inch) long. The clam burrows next to a Urechis and opens its siphon, or water inlet for respiration, into the burrow. This keeps the clam safe, as it can burrow quite deeply and still have access to a fresh flow of water. It lives next to burrows of shrimps,

One of several echiuroid worms known as "fat innkeepers" because "guests" such as pea crabs, fish, and clams share their
burrows. This one was found southeast of Hawaii in the sea
floor. The insert shows it with its proboscis partly retracted
(at right). It was found in part of a sea-floor core, or cylinder
of solid mud removed from the sea bed.

too, and is more frequently found there. Urechis strains
food from the water so efficiently that little is left for
the clam.

One species of scale worm is found only in Urechis
burrows. Only one individual is found in one burrow, however, since a resident worm will fight off any intruders.
The scale worm feeds on particles of food too big for
Urechis to swallow. Its head is always pointing in the

same direction as that of its host, and it always keeps its back touching Urechis. If Urechis turns in the burrow, the scale worm quickly does the same.

## A Miniature Male

One very strange relative of Urechis is named Bonellia. This green worm has a very long proboscis which is forked at the end. A Bonellia only eight centimeters (about three inches) long can extend its proboscis out to a length of a meter (over 39 inches) while hunting for food. But the most peculiar thing about this worm is its way of reproducing. Only the female worm is large. The male is a ciliated creature only three millimeters long. It has no proboscis or circulatory system. It enters the body of the female and lives like a parasite inside her, absorbing food from her body. The sex of most animals is determined when the egg is fertilized. But not Bonellia. The larvas have no sex when they are born into the water. If a larva settles itself in contact with a female, it becomes a male and enters her body. But if it settles down away from other Bonellias, it becomes a female and grows large.

## Horsehair Worms

Anyone who spends much time exploring along streams may well have come across horsehair worms. The long, thin, dark-brown body of this peculiar creature certainly resembles a rather fat hair from a horse's tail. While only about one millimeter across, the body of a horsehair worm is usually about 36 centimeters (over 14 inches) long. The adult worms live in fresh water. Females are quite inactive, but males can swim or crawl around by using whiplike movements. After males and females mate, the eggs are left in the water. When the larvas hatch, they have a spined proboscis which can be pushed out. They enter the bodies of insects such as beetles or grasshoppers which live near the water's edge. There they live, grow, and molt until almost mature. When their hosts are near the water they emerge and soon are able themselves to reproduce.

## The Newest Phylum

The strange deep-sea worms called pogonophorans are a lesson to anyone who thinks we have discovered everything. This phylum was recognized only in the twentieth century. Many scientists had seen pogonophorans without even realizing that they were animals. During the great collecting trips of the British ship *Discovery II* during the 1920s, masses of peculiar fibers were often dredged up from the Antarctic seas. Little else appeared in the dredges. The disgusted and bored biologists on board—many of them top scientists of the time—threw this "useless" material overboard by the ton. Now we know that this annoying junk really was masses of pogonophoran tubes. The biologists had discarded a precious treasure of little-known animals without realizing it.

It is true that at first glance a pogonophoran tube does not look very promising. It seems much too thin to contain an animal. But sure enough—inside that narrow structure lives one of the strangest worms of all.

We have learned of parasites such as tapeworms which

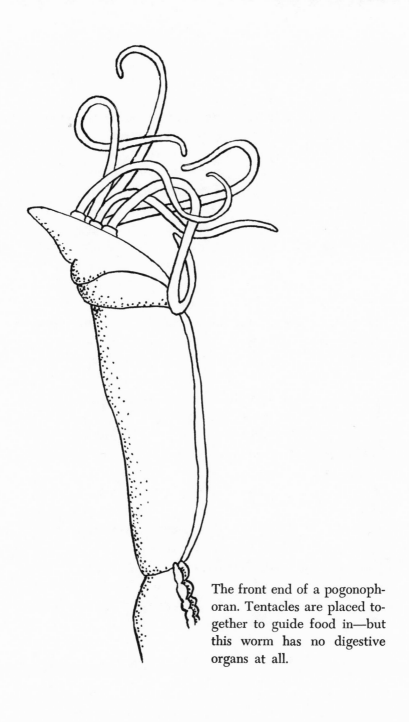

The front end of a pogonophoran. Tentacles are placed together to guide food in—but this worm has no digestive organs at all.

completely lack a digestive system. In a parasite, that is understandable, since there is plenty of nourishment surrounding the animal. But pogonophorans are free-living animals with no digestive system whatsoever. There is no mouth and no gut. Pogonophorans are able to absorb at least some nutrients directly from the water through the body wall, though all the details of their digestion are still unclear. The body is covered with little extensions which give it a huge surface area for absorption.

These worms also have tentacles which can be held in a funnel-shaped array. Cilia pointing inward pull water through this funnel toward the worm. There are gland cells on the tentacles which perhaps secrete enzymes that would break down food particles sticking to the tentacles. The nutrients could then be absorbed directly by the tentacles.

Not only were pogonophorans recognized as animals only in the 1920s, their relationships to other animals were not understood until the 1970s. When pogonophorans are pulled from their muddy bottom homes, the rear portion of the body usually breaks off. This part of the body extends from the bottom of the tube and anchors the animal in the mud. Until this vital section was discovered, scientists thought that pogonophorans were advanced animals. It appeared that they had three body sections, like higher animals. But when the other part of the body was discovered, it became clear that they are probably closely related to annelids. The rear part of the body is segmented and bears setae. And the setae are very similar to those of annelids.

One thing we can learn from pogonophorans—nature

still has surprises in store for us. We must never assume too much. We must always be on the lookout for the unexpected, new, and different secrets which are waiting to be discovered.

# Nine

# Finding Your Own Worms

Now that you have read about so many kinds of worms, you might want to go out and find some for yourself. The easiest worms to find, wherever you live, are roundworms. They live in the plants and soil even in the busiest, most crowded city. The trick is to separate the nematodes from the soil or plants. Nematodes are attracted to water, so they can be lured away from their home by putting it in contact with water. Since they are small, they can pass through a cloth or paper filter which will hold back the other material. Any way you can set up a system in which the sample is separated from some water by a filter of cloth or facial tissue should work.

Here is one method. Place some cloth (try a piece of old sheet or several layers of cheesecloth) in the bottom of a small kitchen strainer. Place the slightly dampened soil or plant material in the strainer. Set this in a paper cup filled with water (it is best to use distilled water) just deeply enough so a sizable area of the strainer screen rests under the water. Wait three or four days, gently adding water if necessary to keep the bottom of the strainer in contact with the water. Then remove the strainer and carefully punch a hole in the bottom of the cup, collecting the first water which comes out to examine with a magnifying glass or microscope. Or else carefully pour off most of the water from the top of the cup and then examine the water from the bottom.

If you want to keep your roundworms for awhile, put them in a glass or enameled pan in a shallow layer of distilled water, uncovered, and they should stay alive for days or even weeks. After some days, when they have cleared out any food in their digestive systems, their other internal structures will be easier to see.

### Flatworms and Fresh-water Oligochaetes

If you have a pond or ditch nearby, gather up some water plants from it and carry them home with a generous amount of pond water. If you can get a gallon jar from a restaurant, it will be useful (after careful rinsing) for carrying and keeping your plant samples. Or use a goldfish bowl. Collect some mud from the bottom, too. If you have a microscope, examine bits of the pond plants with it. You are likely to find oligochaetes like Chaetogaster among them. You may also find little flatworms covered with cilia gliding smoothly along the leaves. One worm called Dalyellia is in some cases bright green. Another one named Microstomum reproduces by budding. You may find chains of these worms in your samples.

Hunt in the mud for Dero, Tubifex, and other oligochaetes. If you leave the mud in the bottom of a jar with pond water until it settles, you may see Tubifex worms waving their tails up out of the mud. If you find them, be sure and keep them in as deep water as possible with plenty of mud, as too much oxygen may kill them. You may find roundworms in the mud, too.

To collect planarians, take along some bits of lean raw beef when you go to the pond or ditch. These worms can also be collected in the quiet parts of clear streams. Put the meat in the water and wait at least fifteen minutes. Then remove it and see if you have managed to attract any planarians. If you have, shake them off into a jar with pond water and discard the meat. Planarians can also be collected from the underside of stones in the water.

You can keep your planarians for a long time if you take good care of them. Put them in a shallow container with some sand, pebbles, water plants, and pond water. Feed them bits of meat such as liver or pieces of earthworms once or twice a week. Always remove uneaten food after an hour or so, and change the water after feeding, using pond water if possible. As second choice, use tap water that has stood in an open container for a day or more, though even with the airing, tap water may disagree with them.

### Leeches

The method for collecting planarians may attract leeches, too. Another way to collect them is to stand in the water of a pond or slow-moving ditch with rubber boots on. Leeches may swim up and attach themselves to your boots, perhaps attracted by whatever bodily warmth comes through them. Simply walk out of the water and carefully remove the leeches by sliding them off with a dull knife.

Leeches can be kept a long time and are very interesting to watch. Be sure to use pond water for them as chemicals in tap water can kill these worms. Make them a home like the one for flatworms, but cover the top with cheesecloth held on with a rubber band, for leeches like to wander. Feed them every few weeks (remember, they can store up lots of food) on bits of raw meat or earthworms. Some leeches will eat snails. Remove leftover food and change the water after each feeding.

If you want to look at a leech with a hand lens or microscope and it is too active, add some soda water or bits of tobacco to the water to slow it down.

## Earthworms

Earthworms are easy to collect simply by digging in the damp soil, in compost heaps, or under damp leaf litter. Or you can go out at night and find them on the soil surface. Take them home and keep them in a container with moist earth. You can feed earthworms many foods— dead leaves, bits of vegetable trimmings, even damp newspaper. Place the food in a slight depression in the middle of your little worm farm and it will magically disappear. If you visit them after dark, you may be able to watch your worms pulling food into their burrows or mating on the surface. Do not surprise them with sudden bright light, however, or they will duck quickly back into the earth. A red light is least likely to be noticed by them.

## Sea Worms

Most of you will not be able to go out and collect sea worms to examine. But if you live in a city with a public aquarium, visit it and examine the rocks in the marine tanks. The chances are you will find some lovely fanworms there. Some public aquariums have displays of Chaetopterus or Urechis living in glass tubes. If there is a tropical fish store in your town which sells marine fish, you may find fanworms on display there, too.

If you live by the sea, you can collect your own worms. Scale worms and nemerteans often live under stones in tide pools. Nemerteans look limp and flabby when you find them. You will know you have one when you pick it up and it shoots out its proboscis, coiling it this way and that in your hand. Fanworms live attached to rocks and shells. One very tiny fanworm called Spirorbis makes white spiral tubes attached to the shells of snails or water plants. If there are beaches near you, examine samples of wet sand near the water's edge for polychaetes. A tidal mudflat can yield various worms such as Arenicola, Chaetopterus, and the clam worm Nereis, which lives in mud or sand. Look for little piles of worm castings or for the chimneys of built-up burrows as clues to where the worms are living. Since sea-water animals are difficult to keep at home, you should probably examine your finds and let them go. You might try keeping tide pool worms for a day or two in a shallow glass or enameled container with sea water, since they are used to unfavorable conditions. Changing to fresh sea water frequently may enable you to keep them for a longer period.

# Glossary

*annelid*: a worm belonging to the phylum Annelida; it has a segmented body and a coelom.

*anus*: the exit of the digestive tract, usually located near the rear of an animal.

*Ascaris*: a large, parasitic roundworm which infects vertebrate hosts.

*asexual*: having no sex; produced without the union of a sperm and egg.

*cercaria*: the fourth, swimming larval stage of a fluke.

*cilia*: microscopic, hairlike projections from the surface of cells, which beat, helping an animal move.

*clitellum*: a swollen region near the front end of a worm such as a leech or earthworm which produces mucus and makes the cocoon for the eggs.

*coelom*: a kind of body cavity found in annelids and other advanced animals which is lined on all sides by a thin layer of cells.

*cuticle*: an outer body layer which is secreted by skin cells and helps protect the body from the environment.

*elytra*: scalelike projections which cover the backs of some polychaete worms.

*epitoke*: the reproductive rear part of some polychaete worms which specially adapts the worm's body for swarming.

*epitoky*: the reproduction method that use epitokes.

*evert*: to turn inside out.

*fluke*: a kind of parasitic flatworm.

*hookworm*: a kind of parasitic roundworm which attaches itself to the wall of the intestine with a circle of hooklike teeth.

*mantle cavity*: the space beneath the mantle, which is a membrane that lines the shell of a snail or other mollusk.

*mesentery*: a sheet of cells which supports the body organs in an animal which has a coelom.

*mucus*: a slimy substance produced by body cells.

*nematode*: a roundworm.

*nemertean*: a ribbonworm.

*nephridia* (*nephridium*, sing.): body organs which collect and get rid of chemical wastes from the body fluids.

*oligochaete*: an annelid, such as an earthworm, which has only a few bristles (setae) and which has a clitellum.

*parapod*: a fleshy projection from the side of the body of a polychaete worm which is stiffened by bristles and which bears bristles. Also called parapodium (parapodia, *pl.*).

*pharynx*: a muscular part of the digestive tract behind the mouth which is used to pump food into the digestive system.

*phylum* (*phyla*, pl.): a group of living things which share a common body plan.

*planarian*: a type of common fresh-water flatworm.

*Platyhelminthes*: flatworms, a phylum of simple worms without an anus or a true head which includes tapeworms, flukes, and free-living worms called turbellarians.

*pogonophoran*: a worm belonging to the phylum Pogonophora which lacks a digestive tract, has at least one tentacle, and is very thin. Pogonophorans live in tubes on the ocean floor.

*polychaete*: a type of annelid worm which has parapods and many setae.

*primary host*: the host of a parasite in which the adult parasites live and reproduce.

*proboscis*: a projection associated with the digestive system which can be turned inside out to help gather food.

*regenerate*: to regrow lost parts of the body.

*ribbon worm*: a type of flattened, elongated worm which has a unique sort of proboscis; the proboscis is completely separate from the digestive tract and can be shot out rapidly to capture prey.

*roundworm*: a worm belonging to the phylum Nematoda; these worms have a straight digestive tract which lies loose within a fluid-filled body space.

*scolex*: the "head" of a tapeworm which has hooks and spines enabling it to hold onto the intestinal lining.

*secondary host*: a host of a parasite which is infected with immature stages.

*septa*: sheets of tissue which separate the bodies of annelid worms into separate segments.

*setae* (*seta*, sing.): the spines or bristles of polychaetes and oligochaetes.

*stolon*: the egg-filled or sperm-filled rear section of some polychaetes that breaks off and carries on reproduction.

*stylet*: a pointed projection at the front end of some roundworms which is used to pierce food.

*tapeworm*: a parasitic flatworm which lives in the intestine of vertebrates. Tapeworms have no digestive system.

*trematode*: another name for fluke.

*trichinosis*: a disease caused by the tiny roundworm called trichina; it is contracted from eating improperly cooked pork (or occasionally other kinds of meat) which contains the young worms.

*trochophore*: the top-shaped larva of polychaete worms.

*turbellarian*: a free-living flatworm.

# Suggested Reading

## Books

Ralph Buchsbaum and Lorus J. Milne, *The Lower Animals: Living Invertebrates of the World* (Doubleday, Garden City, N.Y.)

Philip Goldstein and Margaret Goldstein, *How Parasites Live* (Holiday House, N.Y., 1976)

C. P. Idyll, *Abyss* (3rd rev. ed., T. Y. Crowell, N.Y., 1976)

Marie Jenkins, *Animals Without Parents* (Holiday House, N.Y., 1970)

## Magazine Articles

J. D. George, "Curious Bristleworms: Polychaeta," *Sea Frontiers*, September 1970

Meredith Jones, "Complexities in the Substrate," *Natural History*, May 1963. Tells about marine polychaetes.

———, "Flowers of the Sea," *Natural History*, March 1973. Has lovely photographs of marine fanworms.

D. W. Kirtley, "Reef Builders: Sabellaria," *Natural History*, January 1968

W. H. Nicholas, "Biggest Worm Farm Caters to Platypuses," *National Geographic*, February 1949

J. D. Scott, "Wonder of the Worm," *National Wildlife*, August 1968

Barbara Smelzer, "Night of the Palolo," *Natural History*, November 1969. Mainly folklore associated with the Samoan palolo worm.

R. Spivack, "Featherdusters: Polychaetes, Sabellids and Serpulids," *Sea Frontiers*, November 1976

Kenneth S. Warren, "Precarious Odyssey of an Unconquered Parasite," *Natural History*, May 1974. Tells about a dangerous fluke.

G. P. Wells, "Worm Autobiographies," *Scientific American*, June 1959. Describes studies mainly on Arenicola, the lugworm.

D. P. Willoughby, "Earthworms: At Home Underground," *Science Digest*, November 1968

# Index